Evolution in Plant Design

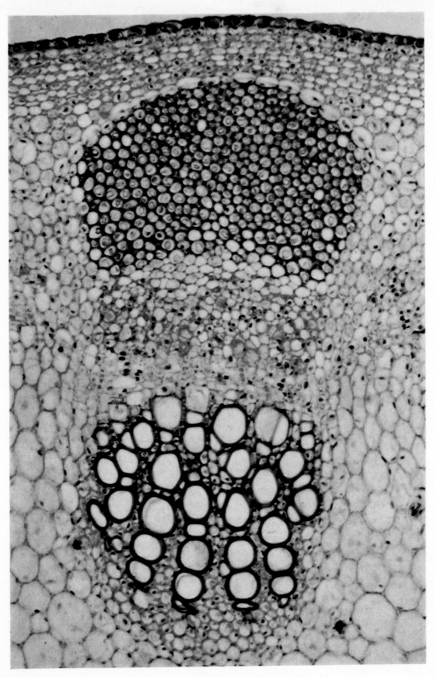

A vascular bundle from the stem of a sunflower

Evolution in Plant Design

C. L. DUDDINGTON

FABER AND FABER LIMITED
LONDON

First published in 1969
by Faber and Faber Limited
24 Russell Square, London W.C.1.
Printed in Great Britain by
W & J Mackay & Co Ltd, Chatham

S.B.N. 571 09065 6

To
LAURENCE D. HILLS
who suggested that I should write this book,
and whose fertile imagination made the
planning of it possible.

Contents

CONTENTS

Illustrations

A vascular bundle from the stem of a sunflower
Photo by Gene Cox, micro colour (International) colour frontispiece

11

TEXT FIGURES

13

ILLUSTRATIONS

Preface

Plants have been in existence for a very long time. Our oldest recognizable fossils date from the Cambrian period, about five hundred million years ago, but it is certain that plant life in some form existed long before that. The first known land plants lived during Silurian times; their fossils are more than three hundred million years old. These early fossils show a comparatively advanced stage, in that they had already developed vascular tissue for the conduction of water up their stems. It is reasonable to suppose, for this and similar reasons, that they were not the first plants to emerge from the sea and inhabit dry land.

During the three hundred million years or so that have passed since the Silurian period, the higher plants have evolved into the varied and complex assemblage that we see around us today. This is the result of natural selection, which has constantly favoured the well adapted, and weeded out the failures. Thanks to this selection, which goes on all the time, today's plants have become beautifully 'designed' for the lives they have to lead.

This book is largely about the higher plants, and the ways in which they are adapted to life in an overcrowded world. It is not intended for botanists only, but rather for the non-scientific reader who is interested in plants—as a gardener, as a naturalist, or just as someone who finds them beautiful. For this reason, I have tried to avoid using scientific terms where simple English would serve equally well. To avoid the use of botanical terminology entirely would have been impossible, but by explaining the meaning of each technical term as it comes along, I hope I have been successful in smoothing the path of the reader without any grounding in science. In Chapter 3 on 'plant hydraulics', as well as in the notes on photosynthesis in Chapter 4, it seemed simpler to use equations than to convey the same information in words only. The glossary, placed for greater

15

usefulness at the beginning of the book, may be useful to those who occasionally have to remind themselves of the meaning of some term that has escaped their memory.

I should like to thank my colleagues at the Polytechnic, Regent Street, for the willing help that they have given me at all stages of the preparation of this book. Without their aid it could never have been written. I should also like to thank Mr. Laurence D. Hills for suggesting in the first place that I should write it, as well as for his invaluable help in working out a preliminary synopsis.

C.L.D.

Kingston-upon-Thames,
1968.

Glossary

In general, terms that are used only once in the text have not been included in the glossary. At their single use they are in quotation marks and accompanied by a brief definition.

Alternation of generations. The regular alternation between a spore-bearing generation, or sporophyte, and a sexual generation, or gametophyte.

Androecium. The male part of a flower, comprising the stamens.

Annulus. A ring of thick-walled cells round the sporangium of a fern.

Angiosperm. A flowering plant, in which the ovules are enclosed in a carpel.

Bract. A leaf subtending an inflorescence or a flower.

Bulb. An underground organ, in which a compressed stem bears a number of scale leaves or leaf bases, containing a bud for the following year's growth.

Bulbil. A bulb-like or tuberous structure arising on the aerial part of a plant.

Calcicole. A plant that thrives on a calcareous soil.

Calcifuge. A plant that thrives on an acid soil.

Callus tissue. Tissue formed as the result of a wound, sealing off the damaged part.

Calyx. The sepals, taken collectively.

Cambium. A layer of cells, usually between the xylem and the phloem of a stem or root, which retains the power of division and growth.

Carotene. A reddish pigment found in chloroplasts and certain kinds of chromoplasts.

Carpel. One of the units composing the gynoecium of a plant, and containing an ovule or ovules.

Casparian strip. The strip of waxy material surrounding a cell of the endodermis.

17

Chamaephyte. A plant life-form in which the aerial part of a plant dies back at the approach of winter, leaving the perennating buds at or near ground level.

Cheiropterophily. Pollination by bats.

Chlorophyll. The green pigment responsible for photosynthesis.

Chloroplast. A solid object in a cell, containing chlorophyll.

Chromoplast. A coloured plastid, or solid cell inclusion.

Chromosome. One of the small bodies in the nucleus of a cell on which the genes that determine the hereditary characteristics of the organism are carried.

Cladode. A leaf-like shoot.

Cleistogamous. Describes a flower that never opens, and is self-pollinated.

Collenchyma. Supporting tissue with unevenly thickened walls that are not lignified.

Column. The joined androecium and gynoecium of an orchid flower.

Companion cell. A parenchyma cell in the phloem of an angiosperm that has a common origin with a sieve tube.

Corm. A compressed, underground stem, serving for perennation and vegetative reproduction.

Corolla. The petals of a flower, taken collectively.

Dichogamy. The condition in which the androecium and gynoecium of a flower mature at different times, thus lessening or avoiding the possibility of self-pollination.

Dicotyledon. A flowering plant that has an embryo with two cotyledons or seed-leaves.

Dioecious. Having male and female reproductive organs on different plants.

Dominant. A plant that dominates a plant community, owing to its tallness, or superior numbers, or both.

Embryo sac. The female gametophyte generation of an angiosperm.

Endodermis. The layer of tissue forming a sheath round the vascular region in a plant root.

Endosperm. The nutritive tissue formed within the embryo sac of seed plants.

Entomophily. Pollination by insects.

Enzyme. An organic catalyst.

Epiphyte. A plant that grows on another one, without being a parasite.

Fibre. An elongated strengthening cell with a thick wall, usually, but not necessarily, lignified.

Filament. The stalk attached to the anther of a stamen.

Gamete. One of the two cells that fuse during sexual reproduction.

Gene. A unit of hereditary material attached to a chromosome.

Geophyte. A plant life form in which the whole of the aerial part of the plant dies away at the approach of winter, leaving the perennating buds below ground level.

Granum. A wafer-like body in a chloroplast, containing chlorophyll.

Gymnosperm. A seed plant in which the ovules are not enclosed in carpels.

Gynoecium. The female part of a flower.

Halophyte. A plant inhabiting a salt marsh.

Helophyte. A marsh plant.

Hemicryptophyte. A plant that dies back to ground level at the onset of the unfavourable season.

Hermaphrodite. A plant that has male and female sex organs in the same flower.

Heterostyly. The condition where a plant has flowers with styles of two or more different lengths, aiding cross-pollination.

Honey guide. A line on the perianth of a flower, guiding an insect towards the nectary.

Hydrophily. Pollination by water.

Hydrophyte. A water plant.

Lenticel. A breathing pore in the periderm of a stem or root.

Mangrove. The name given to a type of plant that grows in a tidal estuary or swamp in the tropics.

Megaspore. The spore that gives rise to the female gametophyte generation.

Malacophily. Pollination by snails.

Micropyle. The opening in an ovule by which the pollen tube enters.

Microspore. The spore that gives rise to the male gametophyte generation.

Monocotyledon. A flowering plant that has an embryo with (usually) one cotyledon or seed-leaf.

Monoecious. Having male and female reproductive organs in different flowers, but on the same plant.

Monopodial growth. Continuation of growth from year to year by the same growing point or bud.

Mycorrhizal association. An association between a fungus and the roots of a higher plant.

19

Nectar. A sugary secretion produced by a plant to attract insects that will effect pollination.

Nectary. A gland that secretes nectar.

Nucellus. The part of an ovule that lies within the integument, and contains the female gametophyte.

Nutation. The twisting movement executed by the growing point of a plant, particularly a climbing plant.

Ornithophily. Pollination by birds.

Ovary. The lower part of the gynoecium of a flower, that contains the ovules.

Ovule. An immature seed.

Parenchyma cell. A cell with living contents, usually thin-walled, rhomboidal in shape, and relatively unspecialized.

Perianth. The outer parts of a flower taken collectively, including calyx and corolla if both are present.

Periderm. Secondary protective tissue developed round the outside of stems and roots.

Peristome. The ring of teeth guarding the top of the capsule of a moss.

Petaloid. Brightly coloured, resembling a petal.

Phanerophyte. Plant, usually woody, where the perennating buds are carried well above ground level.

Phellem. Corky cells formed on the outside of the periderm.

Phelloderm. The inner part of the periderm.

Phellogen (cork cambium). A layer of cells near the surface of the stem or root which, by its growth, produces the periderm.

Phloem. The part of the vascular tissue that is concerned with the conduction of elaborated food materials.

Photoperiodism. The reaction of plants to day length.

Photosynthesis. The manufacture of carbohydrates by green plants from carbon dioxide and water, with the aid of chlorophyll, using light as a source of energy.

Phyllode. An expanded, leaf-like petiole.

Pit. A thin place in a cell wall.

Plumule. The young stem as it grows out of the seed.

Pneumatophore. An erect root provided with aeration tissue, typical of mangroves but also found in a few other plants.

Pollen flower. A flower, without nectar, which is visited by bees for pollen.

Protandry. The condition where the stamens mature before the stigma.

Prothallus. The gametophyte generation in a vascular cryptogam.

Protocorm. A stage in the germination of orchid seeds, before the formation of the seedling.

Protogyny. The condition where the stigma matures before the stamens.

Radicle. The young root as it grows out of the seed.

Receptacle. The tip of a flower stalk, to which the floral parts are attached.

Rhizome. An underground stem, containing reserve food, and serving as an organ of perennation.

Root cap. A mass of cells covering the growing point of a root.

Root hair. A hair, developed near the tip of a root as a simple extension of one of the outer cells of the root.

Rosette plant. A plant having a rosette-like habit, the stem being extremely short.

Sepal. A member of the outer whorl of floral parts. Collectively, the sepals make up the calyx.

Sieve tube. A series of sieve elements in the phloem, arranged end to end, separated by sieve plates. Also used loosely for a sieve element.

Sporangium. A structure in which spores are formed.

Stamen. One of the individual male reproductive organs in a flower, in the upper part (anther) of which the pollen is formed.

Stigma. The receptive surface of the gynoecium of a flower to which the pollen adheres when the flower is pollinated.

Stoma. A pore in the surface of a leaf, surrounded by a pair of guard cells, through which carbon dioxide enters the leaf during photosynthesis.

Stomium. The place in the wall of a fern sporangium where the wall first ruptures when the sporangium opens.

Stroma. The supporting framework of a plastid; in a chloroplast, the colourless portion in which the grana, which contain the chlorophyll, are embedded.

Style. The part of the gynoecium of a flower that connects the ovary and the stigma.

Symbiosis. The phenomenon of two organisms inhabiting one body for mutual benefit.

Sympodial growth. The continuation of growth by a different growing point from that which was responsible for the previous year's growth.

Testa. The tough outer coat of a seed.

Therophyte. An annual plant, relying solely on its seeds for perennation.

21

Tracheid. A conducting element of the wood which has no perforations at its ends.

Transpiration. The evaporation of water from the leaves of a plant.

Tuber. A swollen part of the root, or a swollen underground stem, in which food is stored, and which serves for perennation.

Turgor. The hydrostatic pressure within a cell, resulting from the action of osmotic forces.

Tylose. An outgrowth from a parenchyma cell into the cavity of a tracheid or vessel element, partially or completely blocking the lumen.

Vascular bundle. A strand-like portion of the conducting system in plants, containing xylem and phloem.

Velamen. A layer of empty cells covering the aerial roots of an epiphytic orchid, serving for the rapid absorption of water.

Vessel. A tube-like series of vessel elements, placed end to end.

Vessel element. A cell, usually lignified, of the wood in which the end walls are perforated.

Xanthophyll. A yellow or orange pigment found in the chloroplasts of a plant, and in some chromoplasts.

Xeromorph. A plant showing structural adaptations to conserve water.

Xerophyte. A plant that habitually grows in a dry habitat.

Xylem. The main water-conducting tissue in higher plants.

1 · The problems of plant 'design'

Life began in the sea. We do not know where, and we are not sure when, but most biologists are agreed that it was in the ocean that the first particle of living matter appeared in surroundings that until then had been entirely inorganic. This great event occurred at least two thousand million years ago, and possibly very much earlier.

The first living things were very simple, biologically speaking, though compared with the inorganic molecules around them they were highly complex. We have no idea what they were like, but it is reasonable to suppose that they consisted of protein molecules, possibly not unlike some of the simple viruses that exist today. From these first precursors living cells evolved, probably resembling the microscopic plant-animals that we now call Flagellata.

Somewhere around this point the plant and animal kingdoms diverged and went along their separate paths of evolution. The plants retained the green pigment called chlorophyll, and with it the ability to find their food among the simpler things in Nature's workshop—carbon dioxide, water and a few mineral salts, spun together through the energy of sunlight, made available by chlorophyll. This remarkable process is called photosynthesis. The animals abandoned chlorophyll, and thereafter had to seek organic food, eating plants, or preying on others of their own kind.

In both kingdoms simple one-celled organisms gave rise to more complex organisms in which many cells were joined together. At first the plants were all seaweeds, for the land had yet to be colonized. The sea, however, could not forever hold the constant thrust of evolution. The land was there to be conquered. It was as a land very different from what we see around us today: a waste of barren rocks and deserts, without a blade of grass to be seen, nor even soil, apart from a scattering of dust here and there. None the less it was there; as the fanciful might put it, the land was beckoning to the first plants adventurous enough to leave the hospitable embrace of the ocean and try their luck ashore.

It was during Silurian times, more than three hundred million years ago, that the land received its first plant colonists; even before that plants had probably begun to creep up the foreshore here and there, but not in sufficiently large numbers to count as colonists. At first the invasion must have been very slow. Plants growing near the high tide mark crept gradually landwards until they found themselves growing above the tide, where they were only splashed with spray in rough weather or during exceptionally high spring tides. We can see

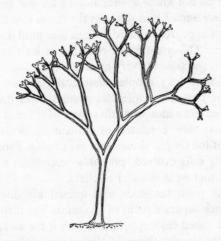

Fig. 1.
The channelled wrack
(*Pelvetia canaliculata*)

the same sort of thing today in the channelled wrack (*Pelvetia canaliculata*), which grows so high up on the shore that it is almost a terrestrial plant (Fig. 1). From here it was only a small step in space to complete divorce from the sea, though a much bigger step in terms of adaptation.

Problems of land plants

As soon as plants left the water they were faced with a number of problems that had to be solved. There was the problem of water supply: the vital processes that go on all the time inside a plant need water, and without it life is impossible. A seaweed has no water problem, for it is bathed in water for at least part of every day. A land plant depends on the water in the soil, and before this becomes available the plant must develop a root system through which the water can be absorbed. A seaweed has no roots, for it does not need them; the most it has is a 'hold-fast' by which it can cling to the rocks, but through which it absorbs nothing. The first land plants had no roots, but they had hair-like organs called 'rhizoids' which penetrated the soil and served for water absorption. Roots were soon developed, however, once plants had taken to the land.

The roots of land plants take in not only water from the soil, but mineral salts as well; the plant depends on these for part of its nutrition. A seaweed again has no problem, for the sea is its source of minerals and the water that bathes it is a kind of nourishing soup that provides all it needs. When plants took to the land they had to find some other way of getting their mineral supply, and their roots became organs for absorbing minerals as well as water.

As land plants increased in stature the problem of lifting water—and with it minerals—from roots to topmost branches became even more acute. The Californian redwood tree can attain three hundred and fifty feet, and the water absorbed by its roots must reach every leaf. Special water-conducting cells were developed to meet this need, and the cohesive force that exists between the water molecules was the means by which water was lifted against the pull of gravity.

A great deal of the organic food needed by the plant is fabricated in the leaves, with the aid of chlorophyll and sunlight. This manufactured food must be distributed to all parts of the plant, down to the tips of the roots. Effective accomplishment of this requires an organized transportation system, in which the cells are adapted to carry amino-acids, sugars and other organic compounds. Such a system is seen in a rudimentary form in some of the larger seaweeds, where special cells form a pipeline along which the various food substances can travel in solution. The land plant, with its body divided into stem, root and leaves, has greater need of such a trans-location system, and a tissue known as the 'phloem' was evolved in answer to this need.

A land plant has to face many mechanical problems that are unknown to a seaweed. The need to stand erect is one. A seaweed is supported by the water. There is no theoretical limit to the size to which a seaweed can grow, and the giant kelp (*Macrocystis*) may attain a length of six hundred feet (Fig. 2). With a land plant it is otherwise. Unless it is going to spend its life trailing upon the ground it must have a stem rigid enough not only to support the branches but also to resist the efforts of the wind to blow it down. This is met by the development of 'xylem', or wood, and by its position in stem and root enabling a stem to resist a 'bending' strain, and enabling a root, which anchors the plant in the ground, to cope with a 'pulling' strain while retaining maximum flexibility.

The earliest land plants had simple cylinders of wood running up the centre of their stems, which served jointly to conduct water and to give rigidity. As they grew in size and complexity, so these central cylinders became split up into smaller units called 'vascular bundles', which came to be arranged in various ways to suit the particular needs of the plant. The nature of the cells that composed the wood also changed with changing needs. Rows of pipe-like cells called 'vessels' were developed, with the sole function of carrying water; other cells called 'fibres', with walls much thickened and strengthened by the

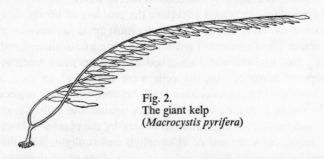

Fig. 2.
The giant kelp
(*Macrocystis pyrifera*)

deposition of a hard substance called 'lignin', took over the job of strengthening the stem and giving it rigidity. Today we make use of these fibres for manufacturing fabrics and cords; linen, for instance, is made from the fibres of the flax (*Linum usitatissimum*) after the soft outer tissues have been rotted off in water.

As plants increased in size and complexity, so the leaves developed as the primary organs in which photosynthesis took place. A seaweed is not divided into stem and leaf, and in the larger seaweeds photosynthesis is carried out by all the cells in the outer part of the flat blade or 'lamina'. The earliest land plants functioned in the same way: the whole of the aerial part was concerned with photosynthesis. With the development of the leaf, a new organ came into being: its function was starch-making. It became flat and thin, so that the cells in which the starch was made got all the light they needed for the process. Leaf mosaics were developed, each leaf fitting into a space left by the leaves above, so that no leaf was shaded more than was absolutely necessary. At the same time, the leaves developed an efficient system of internal ventilation, so that the vital carbon dioxide could enter through openings called 'stomata' and find its way quickly to the photosynthesizing cells.

The ventilation of the leaf, however, had its own dangers for the plant, for if carbon dioxide could get in through the stomatal openings, water vapour could get out. Since the supply of water for a land plant is usually limited, this could lead to death from desiccation. This problem was partly relieved by the development of special cells called 'guard cells' round the stomatal pores, which could close the stomata at night when, owing to lack of light, photosynthesis was not operating.

Perhaps the greatest change of all in the transition from sea to land was in the conditions under which plants had to reproduce themselves. Sexual reproduction involves the fusion of a female egg-cell with a male sperm. For the seaweed, water is always present to carry the sperm to the egg, the sperm usually, though not always, being provided with minute protoplasmic 'tails' called 'flagella' with a waving motion that enables them to swim. All a seaweed has to do is to discharge its sperms into the water, and they will find their way, helped by the movement of the water, to the female egg cells. To make sure of getting a sperm, the egg cells usually emit into the water a chemical which acts as a lure for sperms.

When plants forsook the ocean and came ashore there was no all-embracing sea to carry the sperms, and immediately they found themselves up against a problem in reproduction. This was not solved immediately. For a hundred million years the swimming sperm meant that plants could only reproduce sexually under damp conditions, when the film of moisture covering the plant gave the tiny sperm water in which to swim.

Life story of ferns

Some land plants have never evolved beyond this stage. The ferns, and related plants such as the horsetails and the club-mosses, and also the true mosses and the liverworts, still live with the disadvantage of the swimming sperm. They have adapted to the problem in a curious way, and in so doing have inaugurated a process that eventually led to the higher plants being freed forever from dependence on water for reproduction.

A fern plant does not reproduce sexually at all. Instead, it carries on the underside of its leaflets, or round their edges, thousands of tiny structures called 'sporangia', in which spores are produced. A spore is a minute reproductive structure, rather like a seed but much smaller and much simpler. Its formation needs no fertilization process. A fern spore, given the right conditions of moisture and temperature, can germinate and grow, but in doing so it does not give rise to a new fern plant. Instead, there grows from the spore a plant called a 'prothallus'. This is flat and roughly heart-shaped, and when fully grown it measures no more than a fraction of an inch across.

Few people have noticed fern prothalli, for they lie on the soil, among the litter of fallen leaves. Their small size also makes them inconspicuous. It is on the prothalli that the sexual stage in the life history of the fern takes place, for these little plants bear the male and female sex organs. Close to the ground, with their covering of old leaves and twigs, they have the moist conditions that are essential to the sexual act. The sperms can swim from the male to the female sex organs, and the egg cells are fertilized. From the fertilized egg grows a little plant called—most inappropriately—a 'sporeling', and this, as it develops, becomes a new fern plant (Fig. 3).

Thus, in the life history of the fern, there are two distinct generations: the fern plant proper, which bears the spores, and the prothallus with its sex organs. The two generations follow one another regularly, the fern plant giving rise to the prothallus and the prothallus to the fern plant again. This 'alternation of generations', as it is called, is indelibly printed on the life cycles not only of the fern but of all green plants higher in the evolutionary scale than the seaweeds. It is the answer of the plants to the problem of sexual reproduction on land.

Mosses and liverworts

The alternation of generations shows itself differently in the various groups of plants. In the ferns and other similar plants a prothallus alternates with the larger spore-bearing generation. In the mosses and liverworts, on the other hand, the main plant bears the sex organs. The spore-bearing generation that results from sexual reproduction is the 'capsule', attached to the main plant by a stalk; the capsules of mosses can be seen any day during the spring, rising above the little moss plants on their slender, hair-like supports. The hollow capsule contains spores which, when they are set free to germinate, will

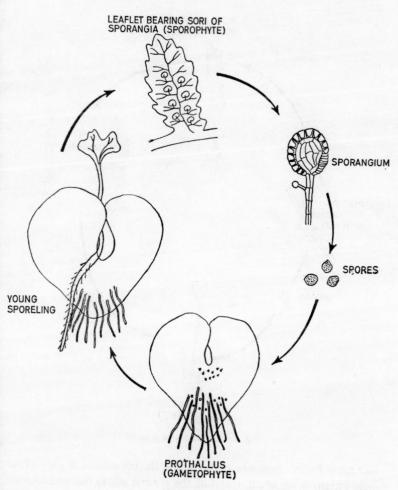

LEAFLET BEARING SORI OF
SPORANGIA (SPOROPHYTE)

SPORANGIUM

SPORES

YOUNG
SPORELING

PROTHALLUS
(GAMETOPHYTE)

Fig. 3. Life history of a fern

produce the moss or liverwort plant. The capsules are, in fact, parasitic on their parent plants, either wholly (as in the liverworts) or partially (as in the mosses) (Fig. 4).

One might ask why the 'main' generation of the moss plant bears sex organs, while the spores are relegated to the capsule. In the ferns it is the other way round: the fern plant produces the spores while the sex organs are borne on the prothallus. There seems to be a difference in emphasis here. The answer is that moss plants are small; even the

29

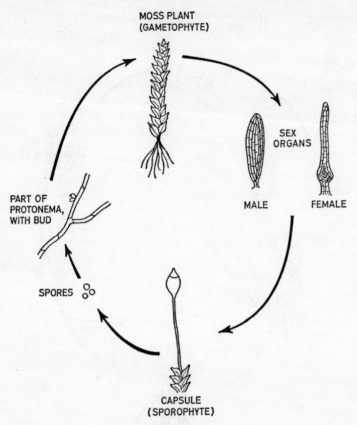

MOSS PLANT
(GAMETOPHYTE)

SEX
ORGANS

MALE   FEMALE

PART OF
PROTONEMA,
WITH BUD

SPORES

CAPSULE
(SPOROPHYTE)

Fig. 4. Life history of a moss

hair moss *Polytrichum*, one of the largest British mosses, is only a few inches high. Being small, they hug the ground where the moisture is. The swimming sperms of the moss do not lack water in which to navigate; there is usually a film of moisture covering a moss plant, at any rate at certain times of the year, and it is no accident that most mosses reproduce in the spring, before the summer sun has dried them up. Liverworts, too, are small. Many of them are flat and ribbon-like, growing in damp places such as streamsides and old culverts, while the leafy liverworts are even smaller than mosses.

If the water problem is not serious for mosses and liverworts, why do they have a spore-bearing generation at all? It would seem simpler to omit the alternation of generations, so that the sexual generation

reproduced itself directly. The reason for the alternation of generations in mosses and liverworts is twofold. In the first place, although
we do not know their ancestry owing to lack of fossil remains, it is
reasonably certain that they are descended from plants in which the
alternation of generations was well established. What has been won
in the hard battleground of evolution tends to persist even when the
need for it is past. There is, however, a more important reason than
that. The presence of spores gives the plant a chance of distribution
that the sexual generation cannot give.

Distribution is generally an easy matter for a seaweed, for the
fertilized egg cells are carried away from the parent plant by the rest-
lessness of the sea. Not so with a land plant. Unless there is a dispersal
mechanism such as we find in the fruits and seeds of flowering plants,
the fertilized egg cell must germinate where it lies after the death of the
parent plant. Dispersal is a crucial matter for plants, since normally
they spend their lives in one spot.

For a spore, dispersal is no problem at all. Spores are small and
light, and can be blown away easily by the wind, sometimes for
hundreds of miles, like the spores of *Puccinia graminis*, the fungus that
causes black rust of wheat. For a land plant the spore is a good agent
of dispersal, while the fertilized egg cell is a bad one. Notice how, in a
moss plant, the spores are held up aloft in their capsules, exposed as
much as possible to the wind. Even the capsule itself has a mechanism
by which the spores are released when the air is dry. The chamber
that contains the spores is shut by a series of teeth—the 'peristome
teeth'—that run from the edge of the capsule inwards. The cells
composing the teeth, which are devoid of living contents, have their
walls thickened in such a way that when the air is dry the teeth open
outwards, releasing the imprisoned spores, while if it is wet they
return to their position guarding the opening of the spore chamber.
Thus, the spores are released only when conditions are most favourable for them to be blown away.

The fern plant, too, raises its sporangia, containing the spores,
aloft on its leaflets, while the prothalli that bear the sex organs hug
the ground where it is moist. The fern also has an arrangement that
ensures that its spores will be scattered in dry weather rather than wet.

The flowering plant shows adaptation to terrestrial life in its
greatest degree of perfection. That is hardly surprising, for it is the
highest product that plant evolution has devised. From the time that
plants first made the perilous migration from the sea on to the land,

two hundred million years passed before the flowering plants appeared. When they did emerge from the cradle of evolution they quickly took possession of the earth.

Alternation of generations in flowering plants

The flowering plant, like the fern that preceded it, has an alternation between a sexual and a spore-bearing generation, but adaptation has gone so far that the two generations are not easy to recognize. The flowering plant itself is the spore-bearing generation. The spores are of two kinds: some (the 'microspores') produce the generation bearing male sex organs, while others (the 'megaspores') give rise to the generation that has female sex organs. The microspores are familiar to everyone, for they are the pollen grains. To recognize the megaspores we need a microscope. Inside the ovary of the flower (the central female part) are the ovules or immature seeds. If we cut a thin slice of an ovule and examine it microscopically we can make out a single large cell called the 'embryo sac'; this is the female sexual generation, and it has developed from a megaspore inside the ovule.

In the flowering plant the sexual generation is very much reduced. There is nothing that can be called a prothallus. When the pollen grain (microspore) germinates it produces a pollen tube that grows into the tissue of the ovary until it enters an ovule. Within the pollen tube are two nuclei, the male sex nuclei or 'gametes'. One of these fuses with the female nucleus or gamete in the embryo sac, and the ovule is fertilized; it can now develop into a seed.

By reducing its sexual generation almost to nothing in this way the flowering plant has become completely independent of water for its sexual process. The swimming sperm has gone, its place being taken by the male nuclei that are brought to the egg nucleus by the growth of the pollen tube. Provided that its roots can find enough water to make life possible, a flowering plant can—and some do—reproduce in the middle of the Sahara desert.

We see in the modern flowering plant an organism that is beautifully adapted to its particular environment and mode of life. It has to be. In the struggle for existence between plant and animal, and plant and plant, there is no room for the ill-adapted. In order to survive, plants have had to adapt themselves, and those that failed to do so have failed to survive, and so have made room for those better fitted to take their place. This is what Darwin meant by the 'survival of the fittest'.

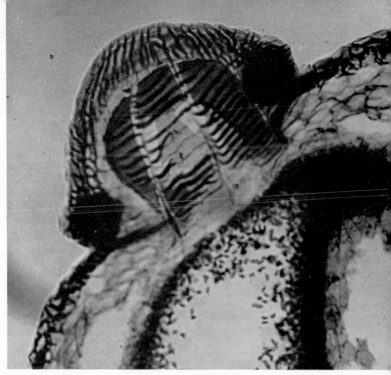

1. Photomicrograph of the top of the capsule of a moss, showing the peristome

2. Photomicrograph of tracheids from the wood of the Scots pine, showing bordered pits in surface view

3. Photomicrograph of tracheids from the wood of the Scots pine, showing bordered pits in sectional view

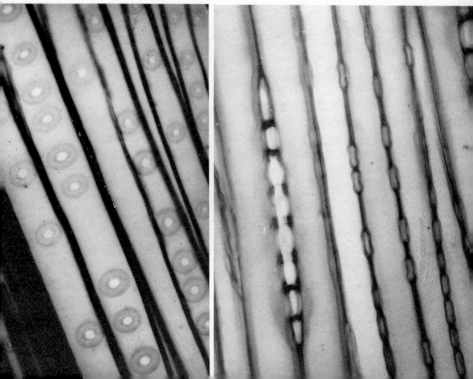

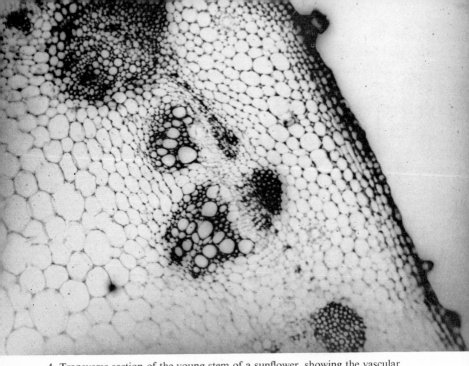

4. Transverse section of the young stem of a sunflower, showing the vascular bundles. The large cells in the bundles are the wood vessels

5. Transverse section of the stem of the dead-nettle. Note the collenchyma in the corner of the stem (jutting out in the photograph)

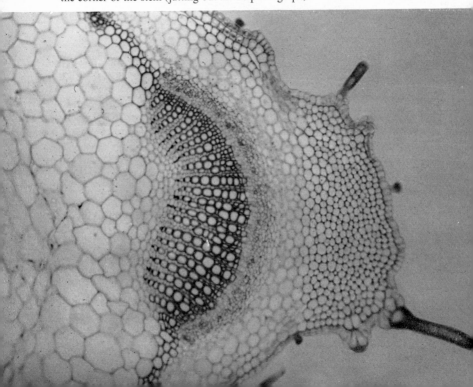

The guiding agency that has watched over this gradual process of evolution, which has been going on since life first appeared on earth, is natural selection. Those organisms which, for any reason, were better fitted to survive the struggle for existence, have lived, and in so doing have passed on their desirable characteristics to their offspring, which in turn have been able to survive. In this way, favourable adaptations have become imprinted on the plants that first produced them, and have been augmented with the passage of the years. Conversely, any individual which showed an unfavourable variation, making it less able to compete with its fellows, has perished, taking with it the deleterious character that killed it. Nature sees to it that only the fit survive.

Natural selection works in much the same way as the artificial selection used by man in building up his domestic breeds of cattle, or his cultivated varieties of crop plants. If a nurseryman comes across a new variety of pea with larger pods or more succulent seeds than usual he uses it to breed more pea plants in which he hopes this character will become established. On the other hand, plants that are of subnormal quality are not used for breeding. Sometimes the new variety arises spontaneously, but not always. Plant and animal breeders may deliberately set out to produce new races by crossing two parents each of which has some desirable quality. When wheat plants which were resistant to the black rust disease were first discovered they were of inferior cropping power, but by crossing them with plants which were good croppers, though prone to rust disease, Professor Biffen of Cambridge produced the first 'Yeoman' wheat, in which the character of rust resistance was combined with the ability to yield heavy crops.

Whether the desirable characteristic arises spontaneously or by the artifices of the breeder, the result is the same: improvement of the variety in question. Man, by using his intelligence, can sometimes—though by no means always—get results quickly. Natural selection, which acts blindly as far as we know, does the same thing, but more slowly. Speed is of no importance to Nature, for she has all the ages to work in.

It is important to note that natural selection cannot by itself *create* new species: it can only guide variations that are already provided for it to work on. That is the supreme importance of sexual reproduction. When two organisms reproduce sexually their characters are blended in the offspring: in some respects it resembles the male parent while in others it takes after the female parent. Some of its characteristics, however, will be purely its own, for the 'genes', as we call the physical stuff of heredity, of the two parents may combine to produce something which is foreign to both.

Anybody who hopefully sows the seed of a Victoria plum is almost certain to be disappointed at the result, for though he may get any one of about fifty different plums, he is almost certain not to get a Victoria. Cross-pollinating insects will have seen to that. If you wish to propagate a Victoria plum, or any of our commercial fruit trees, you must do it by grafting a piece of the original tree on to a suitable stock. Only in that way can you be sure you get what you want. Sexual reproduction is unreliable when it is a matter of preserving the characters of a variety or a species unchanged.

It is these uncertainties attached to the sexual act that provide the material for natural selection to work on; they are the life-blood of evolution, and without them the process would come to a dreary halt. Not only does sexual reproduction provide hybrids; much more important, Nature occasionally throws up a 'sport', or 'mutant' to use the correct term. Natural selection seizes on the mutants and has its way with them.

It may at first seem strange that natural mutants are nearly always deleterious, but there is a good reason for this. When a particular plant—or animal, for that matter, for it applies equally to them—has been in existence for tens of millions of years, it will have become nicely adapted to its mode of life. If it were not, it would not have survived. In other words, the genes that control its heredity will be working in harmony with the environment. A mutation is a change in a gene. In a well-balanced set of genes, this is most likely to upset things. Only occasionally will it result in an improvement. On the rare occasions when it does, there is a favourable mutation for natural selection to work on.

We can see the process of evolution, then, as a gradual change by mutation, the mutants being passed through the sieve of natural selection, those which are beneficial being nurtured, while deleterious mutations—the majority—are rejected. In the sifting operation, the struggle for existence supplies the motive force. Anything that may help the organism in that struggle can look to natural selection to preserve it, while anything harmful is destroyed. That is how all the flowering plants, conifers, ferns, mosses and the rest of the plant kingdom arose during the long ages that have passed since life first made its tentative appearance on this planet. That is how modern plants, in their almost infinite variety and complexity, have evolved from primitive microscopic seaweeds floating in the Pre-Cambrian seas, five hundred million years ago.

In the course of evolution, plants have come to fit their mode of life and the environment that houses them, the guiding hand being natural selection. That natural selection, to the best of our knowledge, works blindly and without thought has made no difference to the result. Plants are designed, sometimes with astonishing accuracy, to fit the niche in life that each one occupies.

2 · Plants as structural engineers

One of the first problems that plants had to solve on emerging from the ocean was that of raising their stems up to the light without the supporting embrace of the sea around them. For a seaweed it is easy. Supported by the water that bathes it on every side, the most it needs for maintaining the buoyancy of its fronds is a few air bladders scattered here and there. The bladders float and hold up the fronds where the light can get at them and carry out its vital role in photosynthesis. Even the giant kelp *Macrocystis* can keep its six hundred foot branches floating gracefully in the ocean currents in this way.

Not so with a land plant. No plant has so far been able to evolve bladders filled with hydrogen that float in the air like barrage balloons, trailing the plant beneath them. It is unlikely that one ever will—though I would hesitate to say that anything is impossible to a plant. A land plant must have built-in rigidity, so that its stem will not only stand erect as a result of its own solid structure, but also resist the efforts of the wind to blow it down.

A very small herbaceous plant can stand erect with the aid of hydrostatic pressure, by blowing its cells out with water. We see the same principle in the inner tube of a bicycle tyre. When deflated it flops; it can be bent into any shape, and is quite incapable of supporting itself. Blow it up with air and it is another story. The tube takes on a circular form and resists strongly any attempt to bend it out of shape. If it is inflated hard enough it is almost as rigid as the wheel itself.

Turgor, rigidity and water conduction

Small plants can do much the same sort of thing. When the cells are distended with water—turgid is the term used by botanists to describe the condition—the plant becomes quite rigid, and any attempt to

36

bend the stem may result in breaking the tissues. Such turgor mechanisms, however, are only suitable for very small plants, or parts of plants, such as the stalks that support the leaves. You can see the system operating—or failing to operate—if you visit an allotment on a frosty winter morning. Observe how the leaves of the Brussels sprout plants are hanging limply instead of standing out stiff and straight. In cold weather the roots of the plant find it difficult to take in water from the soil, so that the turgor of the leaf stalks cannot be maintained and the leaves droop. This is known as 'wilting', and we are all familiar with the way in which shoots of plants picked from the garden wilt on keeping, and how they can be revived by putting them in water.

When a plant gets above a certain size turgor is not adequate to maintain the rigidity of the main stem. For one thing, it would put an intolerable strain on its root system, and on the supply of available water, to maintain turgor throughout all the cells of its stem. Also, as a plant grows taller its weight increases out of proportion to its height and to the strength of its stem. The strength of a plant stem is roughly proportional to its cross-sectional area; make it twice as big and it is four times as strong. But the volume of the plant, on which the weight depends, increases as the cube of its linear dimensions; make it twice as big and it becomes, not four times, but eight times as heavy. As a plant becomes bigger, therefore, it soon reaches a point where no turgor mechanism could possibly keep it rigid.

The answer to the problem of rigidity went hand in hand with the solution of another urgent problem—how to arrange for the conduction of water, containing necessary minerals in solution, from the roots up the stem. The early land plants evolved in such a way that one mechanism served two purposes. Running up their stems was a series of cells called 'tracheids'. These cells were spindle-shaped, and much longer than they were broad (Fig. 5). Their original cell walls, like the walls of most plant cells, were made of cellulose, but inside this was deposited a layer of secondary thickening material in which the already-mentioned hard substance called lignin was prominent. These cells lost their protoplasmic contents as they matured, but the thickened cell walls remained. Their pointed ends interlaced with one another, forming a rigid structure that resisted bending. By virtue of the tracheids that they contained plant stems became rigid structures, able to support the weight of the foliage without recourse to any turgor mechanism.

This was not all. The hollow tracheids running up the stems, without cell-contents, were well suited for the conduction of water. They formed a pipe-line from the bottom of the stem to the top, through which water containing dissolved minerals could travel from the roots to the topmost leaves.

The tracheids of these early plants were the first stage in the development of a specialized tissue called xylem or wood. As plants advanced in evolution, the wood became much more complicated.

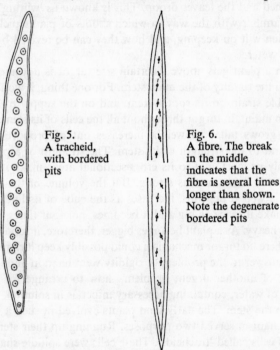

Fig. 5. A tracheid, with bordered pits

Fig. 6. A fibre. The break in the middle indicates that the fibre is several times longer than shown. Note the degenerate bordered pits

As a pipe system the tracheids lacked one thing: each one was shut off from the next by its cell wall, which was impregnated with lignin and should have formed an impenetrable barrier to the passage of water. But the cell walls of the tracheids were not lignified all over. There were thin areas, known as 'pits', where the cell wall remained thin, and through these pits the water was able to pass.

There can be little doubt that the tracheids arose as a dual-purpose element of this kind. Their lignified walls, their interlacing, pointed ends, and the fact that many tracheids occurred together in groups,

38

end to end, running up the stem, gave rigidity. At the same time their emptiness made them obvious candidates for the job of water conduction. The arrangement worked very well—and still does so, for even the giant redwood, the greatest tree of all, depends on tracheids for the conduction of water up its three hundred and fifty-foot stem. Efficient though it was, however, the tracheid could be improved upon.

Because of its water-conducting function the tracheid must retain a certain diameter. If the thickening process were to go on beyond a certain point its internal diameter would become too small to conduct water efficiently—for remember, the thickening is added to the original wall inside, by the living contents of the cell. Moreover, if the wall were to grow too thick, it would become more difficult to arrange for it to be penetrated by the pits that allow the water to pass from tracheid to tracheid. There is a limit to the thickness of a tracheid wall, and therefore to the strength of the individual tracheid.

Fibres and vessel elements

Suppose, however, that a tracheid were to forget about water conduction and just concentrate on strength. It could then lay down as much secondary thickening inside its primary cell wall as it liked, always provided that there was some space left inside it for the living contents responsible for the lignification. We should then have an element far stronger than the conventional tracheid, though of course it would be next door to useless for the conduction of water.

This is just what has happened in the type of plant cell called a fibre. Fibres are long and narrow; almost needle-shaped in some cases (Fig. 6). Their walls are very heavily lignified, leaving only a small central space or lumen where the living protoplasm existed before it finished its work of thickening and died away. Such a cell is specialized for strength and strength alone.

To see fibres at the peak of their development we must go to the flowering plants, though they had begun their evolution in the conifers. The fibres of commerce, used for making textiles or cords, may reach a fantastic length. A single fibre cell of the flax plant may be seventy millimetres long, while those of the ramie (*Boehmeria nivia*), a plant belonging to the nettle family (Urticaceae) and used in the textile industry, may reach the almost incredible length of 250 millimetres.

Fibres are of no value as water carriers, but this does not matter, for there are other tracheids, which have not undergone specialization, that can look after conduction. More important still, in the wood of flowering plants there are cells called 'vessel elements' entirely given over to the carriage of water.

A vessel element has been derived from a tracheid that has become specialized in the opposite direction to a fibre. In the fibre strength is everything; in a vessel element, on the other hand, strength has been sacrificed to water-carrying capacity. These vessel elements have become conductor cells supreme, and they are found almost solely in the flowering plants.

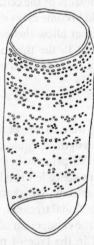

Fig. 7.
A vessel element

The main drawback of a tracheid as a water carrying cell is that the water has to pass from tracheid to tracheid via the pits in the tracheid wall. This it can do, but its passage is slowed down by the need for negotiating so many very narrow openings, each traversed by a pit membrane, the original wall of the tracheid. If only the end wall of the tracheid were not there! The water could then pass from tracheid to tracheid with no more impediment than it would find in a water-main.

This in fact is what the vessel element has done. Imagine two tracheids lying one above the other, their ends joining, connecting with one another through pits in their end walls. The pits become larger, the pit membranes disappear, by mutual enlargement several pits run together, making a single larger hole, and then finally the end walls of the tracheids disappear, leaving free passage between them. Our tracheids have now become vessel elements (Fig. 7).

In the flowering plant we find such vessel elements placed end to end like the segments of a drain pipe. In this way water-conducting tubes are formed, reaching right up the tallest stem. There is nothing to hinder the flow of water up such a system of vessels except the force of gravity that tries to pull it back.

The wider the vessel elements, the more easily the water can flow in them, so it is not surprising to find that, as vessels evolved, the individual elements became wider and wider, and, incidentally, shorter and shorter. We can trace such an evolutionary line running through the families of flowering plants, from the more primitive types of wood to the more advanced. The willows, for instance, have the primitive kind of vessel elements; they are long and comparatively narrow, with tapering ends. In the oak, an advanced form of wood, the vessel elements are short and wide, with their ends flat.

Vessel elements have largely lost their original function of lending rigidity to the stem, their role being water conduction. This does not matter, for the strengthening function has been taken over by fibres, of which there are plenty. Side by side with the vessel elements and fibres, the simple tracheid is also found in many flowering plants. The vessel elements still retain their thickened walls, however. This is necessary, for the deposit of hard lignin prevents the walls from collapsing and so interrupting the supply of water. There is also another important reason why vessels must have rigid walls, but we must defer discussion of that until the next chapter.

Plant stem tissues

A plant stem consists of many different kinds of cell, arranged in a definite pattern that varies between different species of plants, yet keeps more or less to the same general plan. The young stem of a sunflower will serve to illustrate the arrangement of the tissues. The sunflower is an annual plant; the stem, though large, is not designed to persist from year to year, so that the amount of woody tissue in it is small.

If we examine a thin slice or section cut across the sunflower stem we see at once that the bulk of it is made up of thin-walled cells, giving something of the appearance of a honeycomb, though somewhat less regular. These cells have cell-walls made of cellulose and are called by the general term 'parenchyma cells'. They make up what may be termed the ground tissue of the stem. On the outside the stem is bounded by a 'skin' or epidermis, which consists of a single layer of cells.

The ground tissue is divisible into two parts: an outer portion called the cortex, and a central part called the pith. The two parts are separated by a ring of discrete groups of specialized cells, each group being called a 'vascular bundle'—vascular because the bundles contain the cells that conduct water and food materials. It is here that we find the cells of the xylem or wood. Between the vascular bundles the cortex links with the pith (Fig. 8).

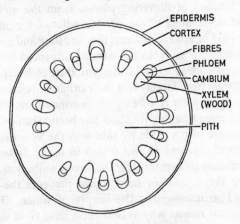

EPIDERMIS
CORTEX
FIBRES
PHLOEM
CAMBIUM
XYLEM (WOOD)
PITH

Fig. 8. Diagram of a transverse section
of a young sunflower stem

The vascular bundles consist of several different kinds of cell, and are all built up according to the same set plan. On the inside is the xylem or wood. This consists mainly of cells with thick walls, and if we dip the section in acidified phloroglucin for a few minutes before examining it under the microscope we find that these cell walls stain bright pink, showing that they contain lignin. The most conspicuous cells in the wood are the vessel elements: cells of large diameter, with walls forming an irregular polygon. The vessel elements are arranged in radiating rows, and between them are smaller cells with thin walls that do not stain with phloroglucin; these are the cells of the wood parenchyma.

Outside the wood the vascular bundle contains two or three layers of thin-walled cells which are flattened, rather like rows of rectangular boxes. These are the 'cambium', which is of great importance when the young stem begins to increase in diameter as it grows. Outside this is the phloem (see page 25 above). This is the tissue especially concerned with the conduction of foodstuffs within the plant, and its functioning will be discussed in Chapter 5. It consists of large, thin-walled cells of a rather empty appearance, called 'sieve tubes', with smaller cells having dense contents scattered amongst them; these are the 'companion cells'. There is also a certain quantitity of phloem parenchyma scattered about amongst the sieve tubes and companion cells. Neither the cambium nor the cells of the phloem stain pink if treated with phloroglucin; their cell walls are not lignified.

Each vascular bundle of the sunflower has a group of cells with very thick walls outside it, like a cap perched on the bundle. The walls of these cells stain bright pink with phloroglucin, showing that they are heavily lignified. These are fibres, and they are the main strengthening elements in the young stems.

The arrangement of the fibres in groups round the periphery of the stem did not occur by chance, for it is here they can be of most use. If you have a certain amount of metal and you want to construct something rigid from it you do not cast it in the form of a solid rod, but rather a tube of greater diameter: the wider the tube, the greater the rigidity, provided that the walls of the tube have a certain minimum strength. This fact is recognized by engineers. It is the material towards the outside that counts; a steel tube, such as is used in a bicycle frame, is nearly as rigid as a solid bar of metal of the same diameter, though of course it is not so strong absolutely.

A plant that has its fibres arranged near the outside of its stem therefore mirrors sound engineering practice. The fibres are placed where they can impart maximum rigidity to the stem without any wastage of material.

This kind of arrangement is seen again and again in the stems of herbaceous plants. The white dead-nettle is a case in point. This plant has a square stem, and if we look at a cross-section under the microscope we see that the vascular bundles are arranged in a square, near the surface. Besides the vascular bundles, the dead-nettle has other patches of strengthening tissue of a different kind. In the corners of the square stem, just beneath the epidermis that covers it, there are groups of cells with walls thickened rather unevenly. Most of the thickening material is concentrated in the cell corners. These cells are not lignified, the added thickening consisting mainly of cellulose; in this respect they differ sharply from fibres, which are nearly always lignified. They also differ from fibres and tracheids in that their living content is retained instead of disappearing when the cells become mature.

This type of strengthening tissue, very common in plants, is called 'collenchyma'. It is found particularly in small plants, and in small parts of larger plants, such as the stalks of leaves and along leaf-blades, where a certain amount of rigidity is wanted. It is apt to occur in seedlings, its strengthening function later being taken over by wood.

The stem of the dead-nettle illustrates another common feature in the construction of herbaceous plants, for it is hollow. This is often the case with herbaceous stems. The central part of the stem is occupied by the pith, which contributes little to the life of the plant. Frequently it is missing altogether from the middle of the stem.

You see this hollow structure particularly well in the stems of many grasses, including that supreme example of plant structural engineering, the bamboo stem. This is one of the hardest of all plant structures, and has even been used by man to make knife-blades.

The bamboos, and grasses generally, belong to the group of flowering plants known as monocotyledons, including also the orchids, lilies, daffodils, irises and other plants with strap-shaped leaves. These usually have their vascular bundles arranged differently in their stems from the single ring that is typical of the dicotyledons, the other great group of flowering plants that includes the sunflower, the dead-nettle and, in fact, most of our common herbs, shrubs and trees.

In the monocotyledons the vascular bundles are scattered throughout the stem instead of being arranged in a single ring. It is so with the bamboo. This has a hollow stem, except at the 'joints' or nodes, and the vascular bundles are concentrated towards the outside. In the cortex, just beneath the epidermis, is a mass of fibres in which the outermost vascular bundles are embedded, thereby imparting great rigidity. This is further enhanced by the walls of the outer cells being impregnated with hard silica, extracted by a piece of typical plant alchemy from the mineral solution absorbed by the roots from the soil and deposited where it will do most good.

The uses to which bamboos are put are very numerous, especially in oriental countries. Bamboo stems combine lightness with hardness, and at the same time are easily split. Entire, they make corner posts for houses, while split they are ready-made tiles for the roof. Walls are made from bamboos that have been split into fine lengths and then woven together to form a mat. Since they are hollow, they furnish segments of pipes to carry water to the dwellings, while split in half they are ideal for gutters. Large bamboos make supports for bridges, while smaller ones provide almost any article under the sun, including walking-sticks, flutes, furniture, household utensils, beehives, agricultural tools and many other things. Split bamboos with their edges sharpened may even be used in grass-cutters. The young shoots of the bamboos may be eaten in much the same way as asparagus, and the stem of *Bambusa arundinacea* contains nodules of silica which are used in Eastern countries as a patent medicine that will 'cure' almost any illness. Nature surpassed herself when she designed the bamboo.

Function of cambium layer

So far we have considered only herbaceous plants, with stems that die away each year on the approach of winter. For a bush or a tree something on a greater scale is required. The increase in size will put far more strain on the rigidity of the main stem or trunk, which will need a massive development of wood if it is to fulfil all its functions. Moreover, the many leaves produced by a tree will place a heavy demand on water supplies, and the wood, as the water carrier, will be involved here also. So it is not surprising that the principal change that we see in the development of the slender stem of the seedling into the trunk of a tree lies in the production of much more wood to cope with the extra work. There are other changes too, but it is the production of extra wood that underlies most of the transformation that takes place.

Let us go back to our original example of the sunflower stem and see what happens when it begins to grow older. As soon as the stem begins to attain any size it undergoes a process called 'secondary growth.' This has its beginning in the cambium—the layer of rather flattened cells that lies between the xylem and phloem in the vascular bundle.

The first thing we notice is that the cells of the cambium begin to divide; one cell divides into two, the two cells then growing to the size of their original mother cell before they in turn may divide again. The cambium of the vascular bundles is starting to form new cells.

At the same time we notice activity beginning in the spaces between the vascular bundles. Level with the cambium of the bundles, cells are also beginning to divide. A cambium is forming in the parenchyma—the tissue between the bundles—which links up with the cambium in the bundles. As a result we soon get a complete ring of cambium, encircling the stem and dividing actively. This is often spoken of as the cambium ring.

We notice two things about the activity of the cambium. One is that the new cells formed as a result of cell division are usually formed radially in the stem—that is, either towards the inside or the outside, but not as a rule between two cambium cells. As the cambium ring gets larger in diameter, as it does with increasing growth, it has to make a certain number of tangential divisions in order that it may accommodate its diameter to the increasing size of the stem, but it makes no more of these than are necessary for this purpose.

The second thing that we notice about the divisions of a cambial cell is that the fate of the two daughter cells is not the same. Usually the outer of the two daughter cells forms a new cell of the cambium, dividing again in due course. The inner cell, on the other hand, ceases to be part of the cambium and differentiates instead into a new cell of the wood—a tracheid, a vessel-element, or possibly a fibre or a wood parenchyma cell. Thus, the cambium is forming new wood all the time, and the new wood is inside the cambium ring.

Sometimes, however, the opposite happens. Instead of the outer daughter cell becoming a new cell of the cambium it is the inner cell that does so. When this happens, the outer cell matures not into a wood cell, but into a phloem cell. Thus, while a cylinder of new wood is being built up inside the cambium ring, a cylinder of new phloem is likewise being developed outside it. The cambium remains in the middle, continuing to divide throughout the life of the plant.

The new wood formed by the cambium is generally similar to the original or primary wood (there are certain differences, but they do not concern us); it is known as 'secondary wood.' Similarly, the phloem formed by the cambium is called the 'secondary phloem.'

We can now see how the massive trunk of a tree is built up. Starting life as a seedling, not so different anatomically from a sunflower, the process of secondary growth forms a great cylinder of wood by virtue of the activity of the cambium. This ever-active layer persists for as long as the tree remains alive, constantly adding more wood as the tree gets older. There is no age when a tree can be said to be fully grown; that is one way in which perennial plants differ from most animals. An animal grows to full size and then stops, but a plant just keeps on growing all its life.

As the internal cylinder of wood gets bigger the girth of the tree increases. The cambium ring would soon get stretched and broken, were it not for the occasional tangential divisions of its cells which increase its circumference in pace with the expanding wood. The secondary phloem, outside the cambium, cannot increase in girth in this way, for once its cells have differentiated as sieve tubes or as something else, that is the end of the road for them. A cell normally can only keep on dividing if it remains simple and unspecialized, like a cambium cell; once it has become anything as complicated as a sieve tube it usually—though not always—loses the power of division. As the wood expands, therefore, it sets up pressure in the secondary phloem, the cells of which become ruptured and crushed. They are replaced as fast as this happens by new phloem cells developed from the cambium. Thus it is that the secondary phloem is always but a thin layer on the outside of the cambium. The secondary wood that makes up the bulk of the trunk of a great tree may be many feet in diameter, but the secondary phloem is paper thin.

The cork cambium

What applies to the secondary phloem applies equally to the layers of tissue outside it—the cortex and the epidermis that in the young plant form a protective covering over the whole. The expanding pressure of the wood disrupts and disorganizes these tissues as the wood grows ever outwards. This might not matter so much, for the cortex of an older plant seldom has much to do, were it not for two things. If the trunk were not covered by some sort of fairly waterproof covering, water could evaporate from it freely into the air, possibly faster than the roots could replace it. No tree could afford to lose water in this way, for water is a most precious commodity to a large plant—or a small one, for that matter. Also, a trunk that was entirely unprotected would be a sitting target for a variety of fungi whose spores are ready to seize on the smallest chink in the outer defences of a tree, such as a wound or even an insect bite. Any good gardener knows that plum trees should only be pruned in summer, when spores of the silver-leaf fungus (*Stereum purpureum*) are least in evidence. This, too, is why we cover the cut branches of trees, after pruning, with wax or paint.

Fortunately, the growth of a new outer covering to replace what is destroyed during secondary growth places no great strain on the ingenuity of a plant, and the process is really very much the same as the activity of the cambium, though less complex. Basically, it consists of the development of a new cambium layer somewhere in the outer tissues of the stem which, by its growth, can keep pace with the increasing girth and form a new outer covering to replace the old one. This new cambium is called the 'cork cambium' or 'phellogen', and the tissue that it forms is the 'periderm'.

The position of the cork cambium varies in different plants. Commonly it is in the cortex, and sometimes it arises just beneath the old epidermis. The cells of the cork cambium divide in the same way as the cells of the vascular cambium. The divisions are radial to the stem, and one daughter cell formed by division remains as part of the cork cambium, undergoing further divisions, while the other daughter cell forms a periderm cell. Its fate depends on whether it is on the outside or the inside of the cork cambium. If it is on the outside, it differentiates to form a cell of the phellem or cork. Its cell wall becomes impregnated with a highly waterproof material called 'suberin', after which, as it gets older, it loses its cell contents and forms part of the rough outer covering of the trunk. If, on the other hand, it is situated on the inside of the cork cambium it forms part of a layer called the 'phelloderm', consisting of rather regular cells whose walls are not suberized and which retain their living contents.

In time the ever-expanding growth of the stem disrupts the cork cambium and the periderm it produces. When this happens, the cork cambium is usually replaced by a new cork cambium which arises deeper in the cortex. This may happen several times, and in the end the cork cambium may arise in the outer layer of the secondary phloem, the primary cortex having completely disappeared.

The periderm of a tree forms part of the bark; the more or less rough exterior of the trunk is, to a large extent, the cork. The bark, however, comprises more than the periderm, for if we strip the bark from a tree we find the naked wood underneath. The bark, therefore, must be everything outside the secondary wood, including the primary cortex (if there is any left), the periderm and the phloem. That is why stripping the bark from the trunk of a tree is so injurious: the bark carries with it the vital phloem, the tissue that is responsible for carrying manufactured food from the leaves, where it is fabricated, to the roots.

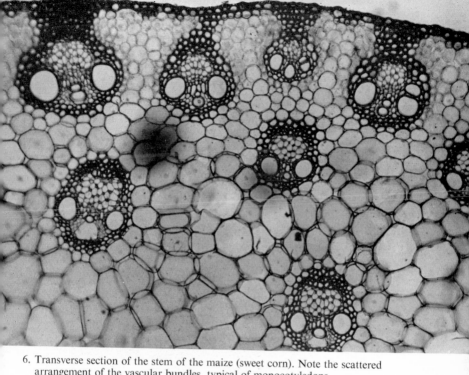

6. Transverse section of the stem of the maize (sweet corn). Note the scattered arrangement of the vascular bundles, typical of monocotyledons

7. Transverse section of the stem of the lime. Note that secondary growth has produced a great deal of secondary wood (bottom of picture)

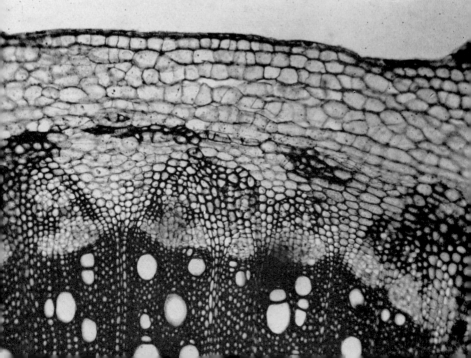

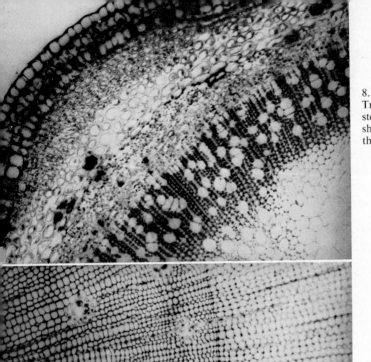

8.
Transverse section of the stem of the currant, showing development of the periderm

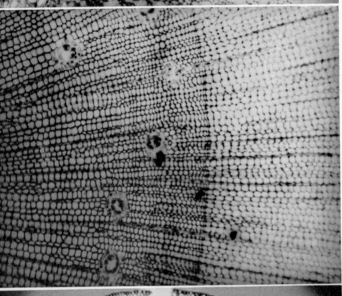

9.
Transverse section through the secondary wood of the Scots pine showing a growth ring

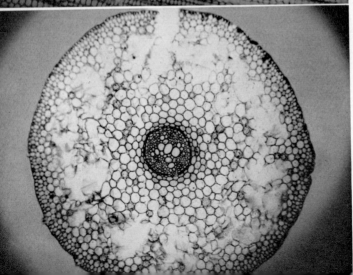

10.
Transverse section of the young root of the broad bean

Since the bark is so important, not only as a protection against infection by fungi but also because of the vital role of the phloem, it is not surprising that a plant has its own means of coping with injury to its bark. If you remove a narrow strip of bark from a tree-trunk you will find that the two cut edges of the wound begin to grow towards one another, finally, if the cut was not too wide, meeting in the centre. This tissue that grows out of a cut edge is called 'callus tissue'; at first its cells are undifferentiated, but they are able to develop into a new cork cambium, and even new phloem cells, to replace what was missing.

The extent of the damage that can be repaired by callus growth in this way is limited. If the bark is removed over too wide an area its replacement may be impossible. It is possible to kill a tree by the removal of a complete ring of bark right round the trunk.

Since the stem contains living cells that are respiring all the time it is necessary for provision to be made for an air supply, for if a stem were covered all over by an impenetrable layer of cork it might be in danger of suffocation. This is provided for by a system of gas exchange pores called 'lenticels'. A lenticel consists of a small patch where the cork cells are missing, being replaced by a loose tissue that allows the ready diffusion of air between them. In this way the need of the living cells for air is catered for, just as the ventilation of the inside of a ship is supplied by air entering through the ventilators. Nothing is forgotten or left to chance in the structural engineering of a plant.

Some trees have very thick periderm. This is especially the case with the cork oak (*Quercus suber*), a Spanish tree, which may have a layer of cork nine inches deep. Bottle cork is obtained by stripping the bark from the cork oak, the strippings being made at intervals with time allowed in between for the tree to recover. The stripping is done at the junction of the periderm with the phloem, so as to cause as little damage as possible, and is a highly skilled job. The dark lines running through a bottle cork are the lenticels in the original cork.

Ageing of tree trunks

The massive trunk of a forest tree contains a vast mass of wood, consisting largely of water-conducting cells, but they do not all carry water all the time. As the trunk grows older the function of water conduction is left to the younger cells of the outer zone of wood, while the inner cells lose their power of carrying water. Various changes take place during this ageing process, including the blockage of the vessels by ingrowths called 'tyloses' (Fig. 9) which are intruded into the vessels from parenchyma cells alongside them. The wood may, in some cases, become darker-coloured as a result of the changes that have taken place, which include the infiltration of the wood with various organic compounds, such as gums, resins, oils, tannins and pigments. Finally, the living parenchyma cells that form part of the wood die.

These changes do not affect the strength of the wood, but they affect its other properties. The altered wood is more durable when cut than the newer wood and more resistant to attack by fungi and bacteria which might set up wood rot. It is also less penetrable by

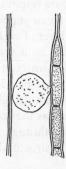

Fig. 9.
Longitudinal section of part of a vessel element, showing closure by a tylose growing out of the neighbouring medullary ray

liquids, and so less easily treated with wood preservatives. The old wood is known commercially as heartwood, as opposed to the wood that is still functioning, which is called sapwood.

In the growing tree the heartwood may be less durable than the sapwood, as is shown by the 'hollow' trees that may be seen growing in woods or by the wayside. Here the heartwood has rotted away, but because the outer sapwood is still alive and functioning the tree is still able to live. This may well be because the sapwood in a living, as distinct from a dead, tree is full of water, and is therefore less liable to attack by a wood-rotting fungus because of poor aeration.

Both the extent of the development of heartwood, and the visible disparities between it and sapwood, differ very much in different timbers. In the poplar, willow, fir and spruce there is no obvious heartwood. Other trees, such as the yew and the mulberry, have a thin layer of sapwood, while in the beech, ash and sycamore the sapwood is thick. The time at which the sapwood becomes heartwood also varies; in some cases the sapwood is easily converted into heart-wood, while in others it functions for a long time before conversion.

Commercial timbers are known as softwoods and hardwoods, according to whether they are derived from a conifer or a broad-leaved tree (flowering plant). This term is a little deceptive, for a softwood is not necessarily soft, nor a hardwood hard. The wood of the pitch pine (*Pinus palustris*), technically a 'softwood', is one of the hardest woods known, while any maker of model aircraft will testify to the softness of the wood of the balsa (*Ochroma lagopus*), a 'hard-wood'. In general, however, conifers do tend to have softer wood than broad-leaved trees. The wood of conifers tends to be simpler and more homogeneous than that of flowering plants (the differences between

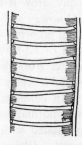

Fig. 10.
Longitudinal section of
part of a tracheid of the yew,
showing the tertiary thickening bands
that give the wood of the yew
its great strength and elasticity

cone-bearing and flower-bearing plants will be explained later, on pages 115 to 116).

The wood of the yew (*Taxus baccata*) is rather interesting because the tracheids which it contains have bands of thickening material (the so-called 'tertiary thickening bands') running round spirally inside the secondary thickening of the cell wall (Fig. 10). This gives the wood of the yew great strength and elasticity—hence its use in archery.

Weight-supporting problems

Plants live under varying loads. Some of these are problems added by the plant breeder, such as too-heavy wheat ears and branches of the Victoria plum which break under the weight of their own fruit. Even without these risks of civilization, however, there are more inherent weight-supporting difficulties to be solved.

Why does the weeping willow weep—or, rather, why do not all

trees weep? A tree grows from a sapling, often without much change of form. The young branches of the sapling tend to jut out at certain angles, and these angles are more or less retained as the tree grows from a sapling into a giant. As more and more wood is added to a branch it naturally gets heavier and heavier. The engineer would regard a branch of a tree as a cantilever loaded with its own weight; as it gets heavier it ought to bend more and more, until finally, if it grows long enough, it would hang vertically downwards under the load. Yet, with few exceptions, this does not happen.

The mechanical stresses that occur in the branch of a tree may be presumed to be the same as in any other cantilever. The material towards the top of the branch is in tension, while the material near the bottom is in compression. If it is to keep a straight course the wood must resist both these stresses equally. How it maintains its apparently predetermined course we do not know.

If we study the wood on the lower sides of branches of coniferous trees, or on the upper sides of branches in broad-leaved trees, we find that it has a distinct structural difference from the rest of the wood. This wood is called 'reaction wood', and it differs from normal wood both anatomically and chemically. In conifers, where it is called 'compression wood', it is darker than normal wood, and the walls of the cells are more heavily impregnated with lignin. The tracheids are shorter than usual, and the wood is denser. The secondary lignified cell wall which is laid down inside the primary wall consists of three layers in normal wood, but in compression wood there are only two layers, the innermost layer missing.

In branches of broad-leaved trees there is no compression wood, but the wood laid down on the upper side of the branch, known as 'tension wood', shows modifications in structure. The vessels are narrower, and the fibres have a thick inner layer—the gelatinous layer—which refracts light very readily and is composed mainly of cellulose instead of lignin.

Recent work has indicated that the formation of reaction tissue is not limited to the wood, for the secondary phloem, the cortex and the cork cambium may also show modified growth.

The study of reaction wood is a relatively new branch of botany, and much remains to be found out before we shall be able to claim to understand its formation fully. Its function, if any, in the life of the plant is at present uncertain, but several prominent workers in this field have expressed the opinion that reaction wood is in some way responsible for the fact that branches do continue growing in the direction in which they start, and do not change their direction of growth as a result of the stresses set up by their own weight. Such a view does explain the position in which reaction wood is found in the plant. If growth regulation is indeed the function of reaction wood, however, it will be very hard indeed to prove it.

The stem of a coniferous tree such as the pine, fir or larch is similar to that of a broad-leaved tree in all essential particulars. Coniferous wood is generally simpler than the wood of broad-leaved trees; for one thing, it contains no vessels, and its fibres are of a primitive type. Secondary growth occurs in the way already described for broad-leaved trees, the activity of a cambium layer adding to the girth of the tree.

Growth rings of trees

It is well known that it is possible to tell the age of a tree, at least approximately, by counting the 'annual rings' that show on the face of the cut stump, allowing one ring for each year of age. These 'annual rings' are more accurately called growth rings, for though normally formed at the rate of one a year, this is not necessarily so.

In our climate, a growth ring is formed by the difference in size between the cells in the spring wood and those of the wood formed the previous summer. After the winter dormancy, life starts off with a burst of energy in the spring. Growth is rapid, and the cells formed are large. As the spring wears into summer there is a general slowing down of growth; the cells formed are smaller. Then comes the winter, and the start of another dormant period, after which there is again a surge of new growth the following spring. There is thus a sharp junction between the spring wood and the previous year's growth, and this expresses itself as a ring seen on the stump when the tree is felled.

The difference between spring and summer wood is largely a function of the differing water requirements of the tree. In the spring growth season, transpiration is high, and large vessels or tracheids are needed to cope with the need for plenty of water. Wood formed later in the year is mainly mechanical in function, adding to the strength of the stem rather than its capacity for carrying water. Hence the elements formed in summer are of smaller diameter.

Growth rings are best seen on wood that grows quickly; the pine, for instance, shows them well. On the other hand, the cycads—curious plants, related to the conifers, often called 'living fossils' because of their ancient lineage and primitive characters—grow so slowly that growth rings are not noticeable.

In these northern latitudes, then, it is the difference between spring and summer wood that produces a growth ring, but this is not true all over the world. In an Indian monsoon forest it is not the cold of winter, but the heat and drought of summer, that slows up growing and gives rise to growth rings. In the tropical rain forest, such as is found in Brazil, West Africa and Sumatra, it is hot and moist all the year round. There is no unfavourable season to slow up growth, and the trees normally do not show growth rings.

In a cross-section of wood the larger vessels show up as holes in the wood, and these are known to the wood expert as 'pores'. We can distinguish two types of wood, according to how the pores are distributed about a growth ring. If the pores are rather uniform in size and scattered fairly evenly about the whole ring the wood is said to be 'diffuse-porous'. Examples are seen in the wood of the sycamore, birch, beech, hornbeam, walnut and poplar. On the other hand, if the pores are very distinctly larger in the spring wood than in the summer wood, the term 'ring-porous' is used. Ring-porous wood is found in the sweet chestnut, ash, oak and elm. On the whole, ring-porous wood seems to be a sign of evolutionary specialization, occurring in comparatively few species, most of them characteristic of the north temperate zone.

Root structure and function

The anatomy of a root differs strikingly from that of a stem. The stresses encountered by a root are quite unlike those a stem has to put up with. A root does not need rigidity; quite the reverse, for the root has to wind its often sinuous way through the soil, avoiding such barriers as rocks and large stones. On the other hand, a root often has to withstand severe pulling strains, as, for instance, when the wind force and direction could lever the stem out of the ground and the root is called upon to provide a firm anchorage. The tip of a stem ends in a bud, but in a root any such delicate structure at the tip would be impossible, for it would soon be damaged beyond repair as the root-tip forced its way through the resistant soil. The tip of a root needs protection for the delicate growing part ('apical meristem') to keep it from being damaged by friction with the soil.

Bearing these things in mind, it is not surprising that we find the woody tissue in a root concentrated at the centre, where it will be most useful. If we look at a cross-section of a young root we find the centre occupied by a solid mass of wood; the mass is star-shaped, with arms radiating towards the periphery. In some roots there is no pith, while in others a central pith is present. The phloem is situated in patches, alternating with the 'arms' of the wood. As in the stem, there is an area of cambium between each patch of phloem and the wood. We find something else, however, that we did not see in the stem. Outside the phloem is a single ring of cells, completely encircling the central mass of wood. These cells are called the 'endodermis', and they can easily be recognized because of the peculiar bands on their walls. These bands contain suberin, and they run right round the cell, as a strap might encircle a box (Fig. 11). The bands are named

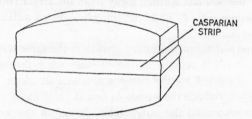

CASPARIAN STRIP

Fig. 11. Diagram of an endodermal cell,
showing the Casparian strip

'Casparian strips', after their discoverer. Their function is a little obscure, but they are thought to have something to do with the movement of water in the root, as we shall see in the next chapter.

Outside the endodermis lies the cortex, consisting usually of thin-walled cells, as in the stem. In a young root the outermost layer of cells forms an epidermis, but as the root gets older this disappears, the outer layer of cells then being known as the 'exodermis'. The cell walls of the exodermis cells contain suberin, and are fairly, though not completely, impervious to water.

Observe how admirably suited a root is for resisting a pulling strain, which is one of its chief functions. The wood concentrated at the centre gives maximum strength combined with the greatest possible flexibility. The structure of the root is like that of a rope or a wire hawser, able to take a load applied to its end, but still capable of being bent round corners. The rigid structure seen in a stem would never do for a root. Once again, plant structural engineering has produced the right form for the job to be done.

I mentioned just now that the delicate growing point of a root might be in danger of being worn away as the tip of the root forces its way through the soil. It is protected from this danger by a structure called the 'root cap'. If we look at a section of a root cut lengthways we can see that the root cap is composed of rather loosely-fitting, thin-walled cells that cover the growing point completely and protect it from harm. As these cells are worn away by the passage of the root through the soil they are replaced by others that grow up beneath them, so that there is always a protective root cap covering the tip of the root.

Although a root has no epidermis, it may develop a periderm in the same way as a stem, especially if exposed to the air. This can be seen where the soil has washed away from the larger roots of a tree with the passage of time. Such roots may be covered with an abundant growth of bark.

A root may undergo secondary growth in the same way as a stem, though some roots remain in the primary state throughout their lives. This is started by the cambium, which develops into a ring round the root, outside the wood. At first the ring has a wavy outline where it passes round the outstretched 'arms' of the wood, but as secondary growth proceeds it grows more quickly in the regions where it is indented, so that it finally becomes a circle. The cambium forms secondary wood inside it, and secondary phloem outside it, just as in the stem.

Besides anchoring the plant firmly in the ground, the root has to absorb water and minerals from the soil. It is well adapted to do this. If we pull a young seedling out of the ground we find that a ball of soil remains stuck to the root and is often impossible to separate from it. This is because there are minute hairs, called 'root hairs', growing out from the root just behind its tip. These hairs are the main site of absorption.

Under the microscope a root hair can be seen to be an outgrowth from a cell at the surface of the root, its cavity being continuous with the cavity of the cell. The nucleus of the root hair cell comes out of the cell and lies in the root hair itself, either near to the tip or about half way along the hair (Fig 12).

The root hairs enormously increase the surface area of the root, and thus make it more efficient as an absorbing organ. There may be two hundred to four hundred hairs per square millimetre of root surface, and it has been calculated that they may increase the surface of the root available for water absorption as much as eighteen times.

Root hairs occur only near the tip of the root, just behind the growing region, and extending backwards for two or three centimetres. Usually the older root hairs cease to function after a time and wither, being replaced by more hairs formed nearer the tip as the root continues to grow. After the root hairs have withered the cells that bore them usually become suberized (corky) and form part of the exodermis of the root; their power of water absorption is then almost (but probably not quite) lost. Sometimes, however, root hairs are more persistent. Many herbaceous members of the daisy family (Compositae) have roots that bear root hairs over most of their length; these hairs may function for as long as three years.

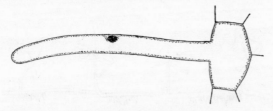

Fig. 12. Diagram of a root hair

When a young seedling is transplanted many of the root hairs are broken and wither away, and those that are not actually broken may wither if they are exposed to the air. That is why it is so important in transplanting seedlings to leave a good ball of soil round the roots; the root hairs will be protected to some extent and some of them may survive. The seedling grows a new set of root hairs after the transplanting operation, and this time of regrowth is rather a critical period, for the water-absorbing power of its roots will be seriously hampered, so that the seedling is liable to die if not kept well watered.

A root hair is much more than a means of increasing the water-absorbing area of the root. It has a very delicate wall, and it grows in intimate association with the soil particles, winding itself round them as closely as possible. The outer layers of its wall are mucilaginous, and merge with the soil colloids so that root hair and soil particle become one structure. This helps the root hair in its work of absorption.

57

Roots have many functions in particular plants besides the primary ones of anchorage and absorption. Often they act as warehouses in which food is stored, the carrot and the parsnip being obvious examples. They may also serve for water storage. This is particularly well seen in the aerial roots of some tropical orchids that grow perched on tree-trunks, sometimes a hundred feet or more from the ground with which their roots make no contact. In some orchids these aerial roots, which hang down like streamers beneath the plant, have a specialized region of cells in their outer cortex called the 'velamen'. The cells of the velamen are large and empty, and can absorb water like a sponge. In this way these 'epiphytic' (growing on other plants) orchids can make the most of such rainfall as comes their way, storing up some of the water against a time of need.

Some climbing plants, such as the ivy, climb by means of adventitious roots on their stems (any root that does not arise in its normal

Fig. 13. Climbing roots of the ivy

place on the end of the stem is called an 'adventitious' root). In the ivy adventitious roots arising on the side of the stem nearest to the wall or tree serving as a support find their way into any cracks or crannies, holding on for dear life and supporting the weight of the ivy plant (Fig. 13). These climbing roots are assisted by the fact that they have a natural tendency to shun the light, growing into any dark spot, such as a crack in a brick wall, that they can find. This sort of behaviour is unusual in roots, which are normally unaffected by light and darkness.

Adventitious roots can of course serve to reproduce a plant, for example when they appear on the cut end of a cutting. An even more dramatic instance can be seen if a cut is made in the surface of a begonia leaf, which is then placed on moist sand and kept covered. After a time, adventitious roots will appear round the cut, and in time a new plant will grow from the leaf.

Sometimes plants have special roots that pull them down into the ground. A crocus grows from a corm, and every year a new corm is formed on top of the old one. As a result of this the crocus would get higher in the soil every year, until finally it appeared on the surface, were it not for the roots that pull it down into position. Besides its normal feeding roots the crocus has special, larger roots that can contract, becoming wrinkled as they do so (Fig. 14). The contraction of these roots pulls the corm lower into the soil, so that it remains at the same depth from year to year.

To see roots reach their full magnificence we must go to the tropics, for the warmth and moisture of a tropical rain forest or a mangrove swamp seem to produce extravagances of growth that are not found elsewhere. In the rain forest 'stilt roots' are quite common, usually on the smaller trees. Stilt roots are adventitious roots that arise on the

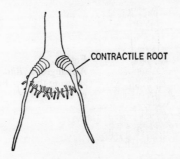

Fig. 14. Crocus corm, showing the
contractile roots

main trunk and, curving downwards, enter the soil. They are very typical of mangroves (Fig. 15), but are seen in many of the trees of the rain forest. They may be round in cross-section or flattened, and occur in trees of many quite unrelated families.

Trees with stilt roots are particularly likely to occur in unstable, swampy soil, and it is tempting to suppose that the stilt roots are there to give extra support. It is always dangerous in biology, however, to accept the apparently obvious explanation without further enquiry. When we look into the matter a little more closely we find that stilt roots usually occur on the smaller trees only, whereas it is the larger trees that really stand in danger of being blown down by the wind. One would imagine that if any trees were to develop stilt roots for support, it would be the taller ones.

A more likely reason for the origin of stilt roots is the warm, moist atmosphere at the lower levels of the tropical rain forest; such conditions would encourage the formation of adventitious roots. This does not mean that they are of no use to the plant, for once formed it would be natural for the plants that had them to take advantage of any support they offered. Plants are very good at seizing on fortuitous peculiarities of structure and turning them to advantage; the story of plant evolution is full of instances where this has taken place.

Even more impressive than stilt roots are the 'buttress' roots that are also found in the rain forest. These are commoner than stilt roots, and the list of families of plants that have them is a very long one. Buttress roots resemble planks leaning, edge uppermost, against the side of the tree. Whereas stilt roots are adventitious roots that

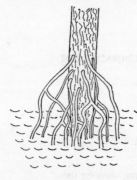

Fig. 15.
A mangrove (*Rhizophora*), showing the development of stilt roots

arise on the trunk and grow downwards, buttresses are developed in connection with lateral roots just beneath the surface of the soil. These produce vertical outgrowths, triangular in profile, that lie against the trunk of the tree.

The wood of buttresses is often harder than the rest of the wood of the same tree, and they tend to be covered with thin bark. If they are examined in section it can be seen from the growth rings that they start as normal circular roots which later develop a strong thickening of the upper side (Fig. 16). The number of buttresses to a trunk may vary from one to ten; most trees so supported have three or more.

Buttressing is mainly characteristic of trees of the tropical rain forest, though it is not confined to them. In general, buttressing is likely to develop where there is a moist, warm climate. It occasionally occurs in temperate forests, especially with the Lombardy poplar (*Populus italica*) and the elm. In the tropics buttressing is particularly notable in trees growing in swamp forests, and on shallow soils with impeded drainage. In lowland forests in the tropics the soil tends to lack aeration, mainly because of the temperature and humidity, and the roots of plants compete with one another for oxygen as well as for water and mineral salts. It is possible that the formation of buttress roots may have something to do with this competition.

The reason for the development of buttress roots is no clearer than the reason for stilt roots. Formerly it was assumed that they were adaptive in that they assisted the tree to stand up to the uprooting action of the wind, especially in soil that was none too firm. This is supported by the fact that one or two buttress roots per tree are

Fig. 16.
Diagram of a buttress root,
cut across to show
the growth rings

relatively uncommon; most trees, as we have just observed, have at least three, the minimum number capable of forming an adequate support. That buttress roots are an advantage in this respect is undeniable, but this does not mean that they have been developed for this purpose. It could equally well be another case of the plants concerned making use of a characteristic that had developed for some other reason.

Various theories have been put forward to explain the formation of buttress roots, but none is entirely satisfactory. One of the most credible is known as the 'strain' theory, which suggests that buttressing is a direct response by the tree to strains produced by the wind. The theory is supported by studies of buttressing in the Lombardy poplar in Switzerland, which tends to develop buttresses on the windward side—the side that has to resist a pulling strain as the wind blows.

Whatever the cause of buttressing, the presence of buttress roots is a constant feature of trees in a rain forest, and the buttresses are often put to commercial use. Those of *Koompassia excelsa*, a Malayan tree, are used for making dining tables.

Another astonishing root development is seen in the 'pneumatophores' or 'breathing roots' that are found in many mangroves. The mangrove is not the name of a plant, but rather of a type of plant that grows in swampy districts such as the Everglades of Florida. Various genera come under the general heading of mangroves, one of the commonest being *Rhizophora*. Mangroves have several noteworthy features, particularly their habit of 'vivipary' or having their young 'born alive' (see Chapter 11 and pages 117, 172); the pneumatophores are just one of these peculiarities.

Living in swamps in tropical regions, a mangrove has difficulty in getting enough oxygen for the respiration of its roots; hence the need for breathing roots to supply the deficiency. The breathing roots are of different kinds. Some are loops rising above the surface of the mud of the swamp, rather like knees, while others are simple branches from the roots that rise upwards, above the surface of the mud, by virtue of the fact that they grow away from the pull of gravity, instead of towards it as most roots do.

Where the breathing roots emerge from the soil into the air they are well supplied with lenticels, and their internal tissues have abundant spaces between the cells by which air can diffuse downwards to supply the buried root system. This lends credibility to the view that they are indeed breathing roots, but they also appear to have another function. In a mangrove swamp the level of the soil is constantly rising by virtue of the litter that is deposited from above, together with the silt that is strained out of the tide by the roots of the mangroves. The main absorbing part of the root system of a mangrove lies in the fine rootlets near to the surface of the soil, and these are formed on the breathing roots, which produce a fresh crop of rootlets every so often at successively higher levels, thus keeping pace with the rising of the soil. In some cases this may be a more important function of the breathing roots than aeration. There will be further discussion of mangroves in Chapter 11, page 192.

Breathing roots are not confined solely to mangroves. The bald cypress (*Taxodium distichum*), a coniferous tree that inhabits temperate and subtropical swamps, has well-developed knee-roots.

The most fantastic development of adventitious roots is seen in the banyan tree (*Ficus benghalensis*). This is a tree belonging to the Fig family (Moraceae). Its squashy fig-like fruits are eaten by bats, and these scatter the seeds. The tree that grows from the seed sends down adventitious roots from its branches which make contact with the soil and develop root systems of their own. In the course of time one single tree can form a small wood, with many 'trunks' that are really adventitious roots. A single banyan tree will allow a small village to nestle beneath its shade. A specimen growing in the Botanic Gardens at Calcutta is a hundred yards in diameter and has over two hundred root 'trunks' in addition to the original trunk, which is twelve feet in diameter. E. J. H. Corner, in his book *The Living Plant*, records that a banyan tree in the Andhra valley measures two thousand feet in circumference, with three hundred and twenty 'trunks' supporting the great weight of its crown.

In India the banyan tree is sacred. Village meetings are held under its shade, and the ground is specially prepared to receive its young roots, which are provided with tubes of bamboo to protect them.

3 · Plant hydraulics

Plants are constantly giving out water from their leaves, particularly during the daytime. Scattered over the surface of their leaves—in most plants, mainly the lower surface—there are innumerable minute openings called 'stomata'. The name is derived from the Greek 'stoma', a mouth, and water vapour issues from the stomata of a leaf as long as they remain open. At night, when it is dark, most stomata close and evaporation of water through them is prevented. Even then, however, the leaf continues to lose water, though at a greatly reduced rate, for a certain amount of water vapour can pass through the epidermis of the leaf instead of through the stomata.

The loss of water from the leaves of a plant is called 'transpiration', and it is surprising how much is lost in this way. It has been calculated that a sunflower plant, in the course of an eighteen-week growing season, may lose as much as six pints of water, while a moderately large birch tree may transpire about ninety gallons on a hot day.

All this water has to be made up by water taken in through the roots from the soil, for if it were not the plant would wilt; and if the wilting were prolonged, death would be certain. No plant can go without water indefinitely, though some, as we shall see, can make do with very little. What goes out must be replaced, and an adequate water intake is one of the first things that a plant must secure for itself if it is to survive.

Osmosis

Before explaining the intake of water by the roots of a plant it will be useful if I remind you of a very simple experiment that you may have performed, or seen carried out, at school. A glass thistle funnel is closed at the broad end by a piece of membrane such as pig's bladder, and, after inverting the funnel, the bulb is filled with a fairly con-

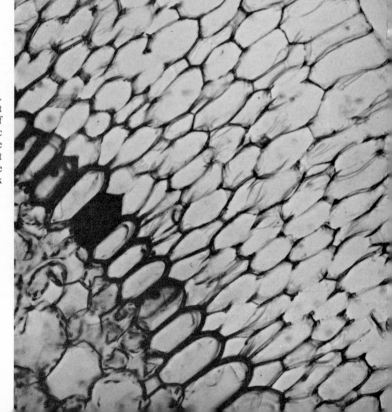

11.
Longitudinal section of the root of an onion, showing apical meristem and root cap

12.
Section through part of the velamen of *Vanda*, an epiphytic orchid. Note the thickened cells that mark the outer edge of the cortex

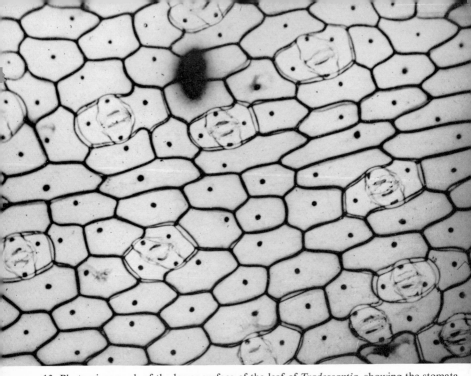

13. Photomicrograph of the lower surface of the leaf of *Tradescantia*, showing the stomata

14. Section through the upper epidermis and palisade layer of a leaf of privet, showing the chloroplasts in the palisade cells

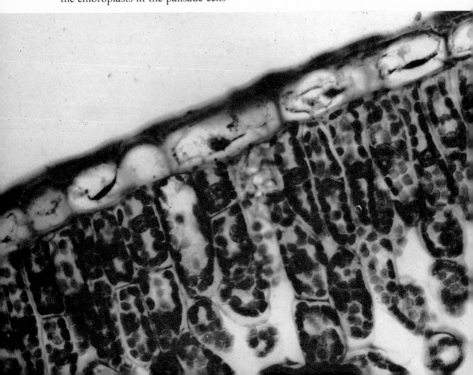

centrated solution of sugar. The bulb is then placed in a beaker of water, supported by a clamp, and left for twenty-four hours. On looking at the funnel on the following day we find that water has passed in through the pig's bladder membrane, causing fluid to rise up the stem of the funnel (Fig. 17). This phenomenon is called 'osmosis'.

The reason for this passage of water is that the membrane of pig's bladder has one peculiar property. It will allow water molecules to go through it quite freely, but will not allow sugar molecules to pass. It is styled a 'semipermeable' membrane, because of this difference in permeability to sugar and water. On one side of the membrane is pure water, the molecules of which can pass through. On the other side are

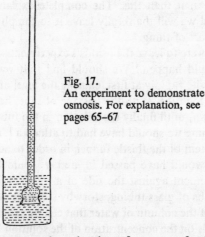

Fig. 17.
An experiment to demonstrate osmosis. For explanation, see pages 65–67

molecules of sugar and molecules of water, of which the water molecules can pass through the membrane while the sugar molecules cannot.

Now the molecules of a liquid, and of a substance dissolved in the liquid, are in a constant state of movement. During the course of this movement water molecules are constantly striking the membrane on both sides, and some of them will pass through it (on one side the membrane will also be struck by sugar molecules, but these cannot pass through). More water molecules will strike the membrane per second on the side where there is pure water than on the side where there is a sugar solution, because on the solution side the water is, as it were, diluted by the sugar. Consequently, over a given period of time, more water will pass into the sugar solution than will pass out of it. Hence as time goes on the sugar solution will accumulate more and more water, becoming diluted as it does so. This is the essence of the osmosis phenomenon. This explanation is over-simplified, for various thermodynamical complications come into the picture. Fortunately, we do not have to worry with them, interesting though they are to a theorist, in order to understand the entry of water into a plant.

I have not attempted to explain *why* the water passes through the semipermeable membrane and sugar does not, and I have no intention of trying to do so. It has been suggested that the membrane acts rather like a molecular sieve, the relatively small water molecules being able to pass through the pores while the larger sugar molecules cannot. This may be true, at least to some extent, but there is almost certainly more in it than this. The complete explanation does not concern us, and we will thankfully leave it to the physical chemists, who like that sort of thing.

Suppose we were to leave the osmosis experiment set up for several days; what would happen? We should find that water entered the thistle funnel quite rapidly at first, causing the level of the solution to rise quickly up the tube. With the passage of time, however, the rate would slow down, until finally the solution in the tube would rise no more. By that time we should have had to attach a long glass tube to the end of the stem of the thistle funnel, in order to accommodate all the water that would have passed in, and it would be advisable to align our experiment against the side of a tall building, so that we could add lengths of glass tubing, story by story, as needed.

The height of the column of water that can be raised by osmosis in this way depends on the concentration of the solution being used, and can be quite fantastic. If our sugar solution contained 342 grams of sugar per litre of water—and this concentration could easily be attained—the height of the column would theoretically be over 750 feet. I say *theoretically*, for various factors would probably conspire to prevent its ever reaching the full theoretical extent—notably, the bursting of the pig's bladder membrane under the enormous pressure set up by the head of water being raised.

This question of pressure is all-important. As soon as the liquid begins to rise in the stem of the thistle funnel it begins to exert a pressure on the membrane, the pressure being measured by the difference in height between the water inside and outside the thistle funnel. A difference in height of thirty-three feet would be equivalent to one atmosphere, or fifteen pounds per square inch. You can see why the membrane would burst if the experiment were carried out for too long a time.

The pressure exerted by the head of liquid rising up the funnel is a measure of what is called the 'osmotic pressure' of the solution. Any solution has a definite osmotic pressure which is proportional to its molecular concentration—that is, to the number of molecules of dissolved substance in a given volume of the solution. One gram-molecule of a substance—its molecular weight in grams—dissolved in a litre of water has an osmotic pressure of 22·4 atmospheres, or about 330 pounds per square inch.

One more thing: I said at the beginning that the bulb of the thistle funnel that contained the sugar solution was immersed in pure water. This might not have been the case, but it would have made no essential difference to the result, provided that the solution in the beaker was lower in molecular concentration than the solution in the thistle funnel. Water would still have passed into the funnel, though less rapidly and not to such a great height. On the other hand, if the solution in the beaker had been of higher molecular concentration than the solution in the funnel, water would have passed *out* of the funnel instead of into it, and the level of liquid in the funnel would have dropped instead of rising.

Let us now see if we can apply all this to the roots of a plant. A root consists mainly of living cells with thin cell walls made of cellulose, a porous substance that allows fluid to pass freely through it. Inside the cell wall is the living protoplasm of the cell, and within this, usually occupying most of the area inside the cell, is a space called the 'vacuole' which is filled with 'cell sap', a complex solution of sugars, organic acids, mineral salts and other things. Somewhere in the cell there is also a nucleus. This may be regarded as a specialized part of the protoplasm; it is denser than the rest of the protoplasm, and appears to exert a controlling influence on the activities of the cell.

Protoplasmic membranes and cell working

The important thing about the entry of water into the cell is the layer of protoplasm that lines the cell wall. It is here that all the reactions occur which give the cell the properties of life—it is, in fact, the essential living part of the cell. Many elements make up the composition of living protoplasm, chief among which are the highly complex organic substances called proteins. The structure of protoplasm is not homogeneous; in particular, studies with the electron miscroscope have shown that its substance is folded into many very fine membranes which play an essential part in the working of the cell.

These membranes appear to be composed of 'phospholipids'—substances of a fatty nature. Two of these membranes are of particular importance in the taking in of water by the cell. One of them, the 'plasma membrane', forms a complete covering to the cell, just inside the cell wall. The other, the 'tonoplast', is on the inside of the protoplasm, separating it from the vacuole. The presence of these membranes, and particularly of the tonoplast, gives living protoplasm the property of being semi-permeable, like the membrane of pig's bladder in our osmosis experiment. A living cell is capable of showing the phenomenon of osmosis and all that it entails.

Let me hasten to add that protoplasm is not truly semi-permeable in the strict sense, for if this were so, it would *ipso facto* be impossible for anything but water to gain entry to the cell, and we know this is not true. But water can enter a cell more easily than most dissolved substances, and to this extent the protoplasm is semi-permeable. Some dissolved substances can enter the cell much more easily than others, but with this we are not immediately concerned. The important thing is that protoplasm is able to act as a semi-permeable membrane in setting up osmotic phenomena.

The cell sap is a moderately concentrated solution, usually with an osmotic pressure of from five to thirty atmospheres. It is bounded by the semi-permeable protoplasm. If the cell is placed in water, or in a dilute solution with an osmotic pressure less than that of the cell sap, water will enter the cell according to the laws of osmosis. This is, in brief, how water goes into the root of a plant.

The water in the soil consists of a very dilute solution of mineral salts of various kinds—chlorides, sulphates, nitrates and so on of various metals such as sodium, potassium, calcium and a number of others. Being a solution it has an osmotic pressure, of course, but that pressure is negligible because its concentration is so small. The osmotic pressure in a root hair cell is much greater than that of the soil solution, so that water enters the cell. Once in the root hair cell it can cross the cortex of the root and enter the tracheids or wood vessels as the case may be. The elements of the wood contain a solution of mineral salts and possibly other things that have a higher osmotic pressure than the soil solution, so that water can enter them from the soil without difficulty.

A formidable number of details still need to be explained, and this simple theory of how water gets into the plant must be amplified considerably before it can win acceptance. To begin with, for the water to cross the cortex from cell to cell there would have to be a gradient of increasing osmotic pressure all the way: each cell that the water entered would have to have a higher osmotic pressure than the cell it left. We have no reason to believe that this is the case. How, then, can we explain the passage of water through the cortex?

Osmotic and turgor pressures

In considering the passage of water into a cell through osmosis I have so far taken no note of the cell wall, since it is made of cellulose which allows water to pass through it without hindrance. Actually, however, the cell wall plays a big part in the process. When water goes into a cell, the cell naturally expands. The cell wall, though apparently rigid, is not quite so; it is slightly elastic and expands a little with the cell; but its powers of expansion are strictly limited, and when forced to expand it exerts a pressure on the cell contents. The more the cell expands, the greater will be this pressure. This inwardly-directed pressure of the cell wall is known as 'turgor pressure'.

We thus have two opposing forces affecting entry of water into the cell: osmotic pressure trying to force water in, and turgor pressure trying to force it out. The relation between osmotic pressure, turgor pressure and water entry can be expressed by the equation

$$P - T = DPD$$

where P is the osmotic pressure of the cell sap, T is the turgor pressure, and DPD is a quantity called the 'diffusion pressure deficit', which is a measure of the tendency for water to enter the cell.

The term 'diffusion pressure deficit' needs a little explanation. Until fairly recently, it was customary to speak of a solution drawing water through a semi-permeable membrane by virtue of its osmotic pressure. This is really putting the cart before the horse. A solution does not *draw* water in. It is more accurate to say that the diffusion pressure of water in a solution is less than that of pure water—or, for that matter, the diffusion pressure of water in a concentrated solution is less than that in a more dilute solution. The more concentrated solution has a diffusion pressure deficit with respect to the more dilute solution. If the two solutions are separated by a semi-permeable membrane, therefore, water will pass from the dilute to the concentrated solution, and will continue to do so until the diffusion pressure of water is the same on both sides of the membrane.

That which determines whether water will enter a cell is not the absolute osmotic pressure of the cell sap, but its DPD. In certain cases this will be the same as the osmotic pressure. In a cell that is about to wilt for lack of water the turgor pressure will clearly be zero, for the cell will be completely flaccid and the wall will have shrunk to its minimum size, so that it does not exert any pressure on the cell contents. From the equation

$$P-T=DPD$$

you can see that, if $T=0$, $DPD=P$.

On the other hand, it is possible for the DPD to be zero even when the osmotic pressure is quite high. As a cell continues to take in water the turgor pressure increases until it becomes equal to the osmotic pressure of the cell sap. At this point no more water will enter the cell, which is said to be fully turgid. The DPD will have become zero. Again, we can see from the equation

$$P-T=DPD$$

that, if $T=P$, $DPD=0$.

For the sake of completeness I might just mention one other case that is sometimes seen, especially in experimental work. Suppose the solution outside the cell is more concentrated than the cell sap. Water will then pass out of the cell instead of into it. The cell wall will shrink to its minimum size and then stop. The protoplasm, however, is non-rigid, so that it will go on shrinking after the cell wall has stopped, losing contact with the cell wall and assuming a rounded or oval shape. A cell in this state is said to be 'plasmolysed'.

From all this it will be seen that the important thing in deciding whether water will enter a cell is not the absolute osmotic pressure of the cell, but the DPD—the difference between the osmotic pressure and the wall pressure. A cell might have quite a high osmotic pressure, but if the wall pressure were correspondingly great, its DPD might be quite small. A cell with a lower osmotic pressure, on the other hand, might have a higher DPD if its wall pressure were small.

There is therefore no need to have a gradient of increasing osmotic pressures from cell to cell for water to cross the cortex; what we really need is a gradient of increasing DPD. If this gradient existed, water would travel across the cortex without any difficulty.

Unfortunately, attempts to demonstrate this DPD gradient have failed, but this does not mean that it does not exist. Furthermore, cells which are surrounded by other cells, as for instance in a root, will interact on one another. If a cell in a tissue begins to swell, it exerts a pressure on the cells that surround it, which may also be trying to expand themselves. In this way, a tissue tension is set up, which will augment the effect of simple turgor pressure.

That tissues may be under tension may readily be seen by anyone who makes a lengthwise cut in the stalk of a dandelion. The two halves of the stalk immediately separate at the cut end and curl outwards. This is because the outer tissues of the stalk are relatively inextensible, so that the central cells are kept under pressure. When the pressure is removed by slitting the stalk the inner cells expand, causing the two halves to bend.

The exact details of how the water crosses the cortex, then, are still a little uncertain, though one may safely hazard a guess that osmotic pressure plays a big part in it. Yet we have reason to believe that osmotic pressure is not the only force at work, and evidence that something else may be operating as well, though exactly what this extra force is we do not know. It is something to do with the fact that the cell is alive, and is therefore known as 'active water uptake' to show that it is bound up with the vital activity of the cell. Possibly it depends on changes in electric charges in a cell on the two sides of the cell membrane.

If this active water uptake really exists, it must need energy to keep it going (as opposed to simple osmotic forces in which the energy is the kinetic energy of the molecules themselves, which needs no augmentation from outside). This energy would have to be supplied by respiration, and it is interesting to note that in some cases the rate of movement of water in plant tissues has actually been found to depend on respiration.

I hope the reader does not feel that he has been led up the garden by my talking at such length about osmosis and then confessing that we do not know exactly how water manages to cross the cortex of a root. Osmosis plays at least a major part in it, if not the whole part; it is only the details about which we are uncertain.

When water has succeeded in crossing the root cortex it has to enter the main water-conducting system of the plant—the tracheids and vessels of the wood. This does not seem to present much theoretical difficulty. The leaves of the plant are constantly giving off water during transpiration, and the water is supplied by the wood. As the water is withdrawn a condition of water tension is set up in the wood supplying the leaves, and this tension is transmitted down to the root. Water is therefore drawn into the wood to replace that transpired by the leaves.

The existence of this state of tension in the wood can be shown if a plant is allowed to wilt and its stem is then cut through under water containing a little of the red dye called eosin. The eosin will be rapidly drawn into the wood, which will be stained red for a considerable distance above and below the cut.

The fluid in the vessels and tracheids is usually not pure water, but water containing various things in solution. Its osmotic pressure is normally greater than that of the soil solution, so that it will tend to take in water from the soil by osmosis, the outer tissues of the root acting as a semi-permeable membrane. It must be remembered that the lignified walls of the wood elements are fairly rigid, so that the wall pressure in them will be small or absent altogether. Although the osmotic pressure of the solution they contain may not be very high in absolute terms, their DPD may well be higher than that of the cortical cells with all the turgor pressure and tissue pressure that they have to cope with. It seems likely, therefore, that entry of water into the wood will be assisted by osmosis.

Role of the endodermis

There is one tissue that I have not so far mentioned in connection with water movement in the root; this is the endodermis. You will remember that it is a sheath of cells that surrounds the wood, shutting it off from the cortex, and that the walls of the endodermal cells have those curious bands, the Casparian strips, around them. You find an endodermis well developed in all roots, but seldom in stems of land plants, though the sunflower stem is one of the few that show signs of an abortive endodermis. Submerged stems of water plants, on the other hand, usually have a distinct endodermis.

For a long time the function of the endodermis in roots was completely unknown, and even now we have not got beyond the stage of theorizing about it. Its constant presence in roots and absence from stems does suggest, however, that it may have something to do with the passage of water. The Casparian strips of the endodermal cells are well larded with fatty materials, which must make them highly waterproof, except for a few cells here and there called 'passage cells', which are free from fat. This may be the clue to the function of the endodermis.

We do not know for certain whether the water that crosses the cortex travels through the living cells or not, for it could pass between them, in the cell walls. Remember, the cell walls are made of cellulose, which is freely permeable to water. When the water reaches the endodermis, however, it finds its way through the cell walls barred by the fatty substances they contain, and is therefore forced to pass through the living cell contents. This immediately brings it under osmotic control.

Whether this is the true explanation of why roots have an endodermis is not known, nor do we understand exactly what advantage the plant would gain by having its water intake controlled in this way. Yet it does seem to be the most feasible theory so far of the function of the endodermis, and so it will have to do until someone can think of a better one.

Having reached the wood, the water passes up the stem to the leaves (never mind how for the moment). From the main wood of the stem strands of xylem and phloem, called 'leaf traces', run out through the cortex, along the leaf stalks, and into the leaves, where they form the veins that are such a conspicuous feature of almost any leaf. In the leaf the veins get smaller and smaller as they branch, and the wood that they carry finally dwindles to a single vessel or tracheid before coming to an end in the tissue of the leaf. These bundle 'endings', as they are called, are so numerous that no leaf cell is very far away from one of them. In this way the precious water is brought to every cell of the leaf.

Leaf structure and function

Leaves vary very much in the details of their structure, but most of them adhere to the same general plan. On top the leaf is covered with the usual 'skin' or epidermis, consisting of a single layer of cells which fit together rather closely; seen from above they look like the pieces of a jig-saw puzzle. The outer surface of the epidermis is covered with a waxy material, forming the cuticle, which is moderately impervious to water; the presence of the cuticle helps to check undue evaporation of water from the leaf.

Below the upper epidermis we find a layer of large rectangular cells, regularly placed with their long axes perpendicular to the epidermis. These cells are called the 'palisade', from their obvious resemblance to a fence of pales. The palisade cells are packed with large numbers of tiny green discs known as 'chloroplasts'; these are the structures that contain the green chlorophyll on which the plant depends for its food. It is the presence of chloroplasts in its cells that makes a plant appear green.

Below the palisade again we have several layers of cells rather irregular in shape and loosely packed together with plenty of air spaces between them. These cells contain chloroplasts, though not so many as are found in the palisade cells; they are called the 'spongy mesophyll', from their loose texture and their position in the middle of the leaf.

Finally, below the spongy mesophyll we have the lower epidermis. This is in general similar to the upper epidermis, and has a cuticle outside it, but it possesses one feature that the upper epidermis often lacks. The lower epidermis is perforated all over by tiny openings, called stomata. The stomatal opening lies between two special cells called 'guard cells', which, in surface view, are kidney-shaped. If a section is cut through the guard cells they can be seen to have walls that are thickened in a peculiar way. The outer walls remain thin, while the inner walls are thickened unevenly. As a result of the rigidity of the thickened parts of the walls and the elasticity of the thin parts the guard cells, when turgid (full of water), swell away from one another, so that the pore of the stoma is opened. When the guard cells lose water they collapse together, thus closing the pore. This means that if the leaf contains plenty of water the stomata tend to open, allowing water to evaporate from inside the leaf; if, on the other hand, water is scarce, they tend to close, thereby cutting down transpiration.

Stated baldly like this it sounds as if the stomata with their guard cells were an almost perfect mechanism for adjusting the amount of transpiration to the needs of the leaf, but this in fact is not so. Stomata are anything but perfect water adjusters, as we shall see in a moment.

In considering the water relations of the leaf it is important to remember that the function of a leaf is the manufacture of organic food for the plant out of carbon dioxide in the air, and water, by photosynthesis. The leaves feed the plant, and the means whereby they do this will be the subject of the next chapter. This function is all-important; if it were not carried out the plant would starve. If this simple fact is borne in mind the somewhat peculiar behaviour of the leaf with regard to water is explicable.

The mesophyll cells of the leaf contain substances in solution in their cell sap, and water passes into them by osmosis from the bundle endings. The leaf cells are turgid, and their cell walls are kept moist. Water evaporates from the moist cell walls into the air spaces between the mesophyll cells. The air spaces link up with one another and with the spaces that occur above each stoma. Water vapour diffuses along the air spaces and finally emerges, through the stomata, into the outer air. Some of it also passes through the cuticle of the epidermis, but the amount of this 'cuticular transpiration', as it is called, is small compared with the amount passing out of the stomata.

The actual area of the combined stomatal pores is small when compared with the total area of the surface of the leaf; it amounts, usually, to about two or three per cent of the whole. One would expect, therefore, that the amount of water that could get away through the stomata would be relatively small. In fact, this is not so. Measurements have shown that the relative transpiration of a leaf— that is, the amount of water that evaporates from a leaf compared with the amount that evaporates from a free water surface of the same area—is in the neighbourhood of ninety-five per cent. In other words, water can get away through the stomata almost as easily as if they and the epidermis were not there at all and the water were evaporated from the wet mesophyll cells straight into the surrounding air.

This somewhat surprising fact is explained when we study the passage of vapours through a partition perforated by a number of small holes, as was done by Brown and Escombe. We find that the molecules do not travel straight when they get near the partition; instead, they bunch together in the neighbourhood of the holes, fanning out again when they have passed through (Fig. 18). As a result, they get through more quickly than one would expect. In scientific language, we say that the molecules of the vapour form 'shells of diffusion' in the neighbourhood of the holes.

A useful analogy can be seen if one studies the crowd entering Twickenham Rugby Football ground, about half an hour or so before the start of the University match. On the east side of the ground there is a broad promenade which runs the length of the east stand. Across it is a row of turnstiles. People are entering the ground all the time, forming a fairly dense crowd but not actually jostling one another, as there is plenty of room. Near the turnstiles, however, they have to bunch up to get through. Those whose path leads them directly to a turnstile pass through relatively unimpeded, while those who are not headed for a turnstile have to push their way to right or to left to get in. The farther they are from a turnstile, the harder they have to push. Once through, they can fan out again and resume their relatively orderly progress.

If an observer were watching the crowd from an aircraft he might get a slightly different view of the proceedings. He would see no details of the horseplay round the turnstiles. To him the crowd would be passing into the ground, through the turnstiles, in a regular stream, apparently unimpeded by the barrier, though he would of course notice that in the neighbourhood of the turnstiles people appeared to

Fig. 18. Path followed by a vapour diffusing
through a membrane with small pores

be bunched together. He would put down in his notebook that a row of turnstiles was practically no barrier at all to the passage of a crowd of people—and in one sense, at any rate, he would be right.

Ascent of sap in stems and trunks

We have seen how water gets into a plant, and how it gets out again, but we have not yet asked ourselves how it rises up the stem into the leaves. This has always been a thorny question for the plant physiologist, and though we have now found a theory that will explain the facts adequately, we cannot claim complete certainty.

The difficulty lies in the great height to which water has to be raised in a large tree. Heights of over a hundred feet are common even in this country, and when we consider the Californian redwood, with its stature well over three hundred feet, and the Australian eucalyptus or blue gum, which is nearly as tall, the problem becomes truly daunting.

There are three possibilities from which we have to choose. The water may be pushed up from the roots, either by osmotic pressure, 'capillarity' (explained shortly) or some other means. Alternatively, it may be sucked up by the leaves, osmosis again supplying the motive power. There is a third possibility: the water might be pumped up the stem by stages, the living cells of the wood somehow acting as pumping stations. Let us take a closer look at each of these possibilities.

The idea that water is sucked up from above seems to be ruled out straight away, at any rate for a tree taller than thirty-three feet, for this is the height of the 'water barometer'. Suction depends on the pressure of the atmosphere to push the column of fluid up. When we suck at the end of a tube we create a partial vacuum at the top, which the air rushes in to fill. If the lower end of the tube is under water, the pressure of the air on the surface of the water causes water to rise up the tube to fill the vacant space. But even if the vacuum at the top of the tube were a perfect one, the pressure of the atmosphere can only support a column of water of a certain height. When the column of water is about thirty-three feet high its weight just balances the atmospheric pressure, after which equilibrium will be established and the water will rise no higher. Hence it would be quite impossible to suck water to the top of even a moderate-sized oak, let alone a redwood.

Looking now at the idea that the water might be pushed up from below, we find two possible mechanisms that could account for its rise. One is capillarity. If the end of a very fine tube is pushed below the surface of water, the water rises up the tube, the phenomenon being known as capillarity. The finer the tube, the higher the capillary rise. The tracheids and vessels of the wood are very fine tubes, and so the phenomenon of capillarity would apply to them just as to any other tube. Here, then, is a possible mechanism that could account for the rise of water up the stem of a plant.

Unfortunately, this idea must be discarded. Although the wood elements are very small, and the minute pores that have been shown to exist in their walls are even finer, capillarity could not account adequately for the ascent of water up the trunk of a tree. Capillary forces are just not strong enough to raise water to a sufficient height.

The other mechanism that might account for the rise of water up a stem is the phenomenon called 'root pressure'. If the stem of a plant that is growing vigorously is cut off at or near ground level, water can be seen to issue from the cut stump. If a tube containing mercury be attached to the stump the mercury will be pushed up the tube, showing that the water is issuing under pressure. The height to which the mercury rises in the mercury manometer, as the tube is called, will be a measure of the pressure that is developed.

Root pressure has been known for a long time. It was discovered by Stephen Hales in 1727, and Hales, by attaching a manometer tube to the cut stump of a stem of the grape vine (*Vitis vinifera*), found that the plant developed a root pressure of one atmosphere. Other measurements have given even greater values for root pressure in various plants; the tomato (*Lycopersicon esculentum*) has recently been found to develop a root pressure of eight to ten atmospheres.

The available evidence seems to indicate that root pressure is an osmotic phenomenon. If the roots of the plant under investigation are immersed in a solution of osmotic pressure equal to that in the wood vessels the root pressure disappears. This is what we would expect if the motive power were osmosis. On the other hand, there seems to be more to it than simple osmosis. One would imagine that water being drawn into the vessels would soon wash them clear of dissolved substances, in which case the root pressure would fade away to nothing quite quickly; but it does not. It looks as if the living cells associated with the vessels must be sending solutes into them as fast as they are washed out.

Some workers have suggested that root pressure depends on the vital activity of the living cells of the root rather than simple osmosis, pointing out that it is adversely affected both by lack of oxygen and by the introduction of poisons into the root. This argument is a fallacy, however. If the maintenance of root pressure depends on living cells secreting dissolved substances into the wood vessels, it is this activity rather than the actual root pressure that would be affected by lack of oxygen or by poisons.

At first sight it would seem as if in root pressure we have the answer to our problem, but when we go into it more fully we find that root pressure, like capillarity, is inadequate. In a small plant there is no difficulty; root pressure, assisted by suction from the leaves, could easily account for the movement of water through the plant. When we come to a tree, however, it is different. For one thing, root pressure seems to be absent in conifers, and yet the conifers are the tallest of the trees. Also, root pressure varies very much with the state of the plant. Large root pressures are shown only by actively growing plants, and at times the root pressure of a plant may even have a negative value. In such cases, water would be drawn down the stem instead of up it. Root pressure cannot provide us with the force we want.

Our third possibility was that living cells in the trunk of a tree might provide a series of pumping stations by which water is raised in stages. This is an attractive theory, but the facts are against it. It was shown by Strasburger and others at the turn of the century that poisonous substances such as picric acid, introduced into the trunk of a tree through cut vessels, rose to a height of many feet and even reached the leaves, although they would have killed any living cells they met on the way. At about the same time, it was shown that when a living tree trunk was placed horizontally, water moved with equal speed in both directions—up the stem towards the leaves and down in the direction of the roots. It was clear, therefore, that the stem contained nothing in the form of valves to ensure that the water flowed in one direction only, as there must surely have been if the rise of water in a stem depends on the pumping action of living cells.

We seem to have reached an impasse. Water is not sucked up from the leaves, nor is it hoisted from below by root pressure, except, perhaps, in small plants. Our third possibility, that it is pumped up by some intermediate cell mechanism, also appears to be without foundation. One is tempted to say that it gets up by the grace of God, but scientists are chary about calling upon divine intervention in order to explain awkward facts. We must look again at our various possibilities (or impossibilities) and see whether we have missed something.

Turning once more to the theory that water is raised from above we can see at once that there is plenty of force available for the job, for transpiration sets up an osmotic gradient in the mesophyll of the leaf with pressures well above the ten or twelve atmospheres that we need for raising water even to the top of a *Sequoia*. The trouble begins when we try to see how it is applied. Suction just will not do, for we cannot get around the fact that the water barometer cannot exceed thirty-three feet. Some other phenomenon must be involved; something that can enable the pull in the leaves to reach right down into the roots and haul the water up. The answer to the problem seems to lie in a theory first put forward some time ago by Dixon and Joly, known as the 'cohesion theory'. A solid substance is held together by forces of cohesion between its molecules, which form a rigid crystal lattice that resists being broken. In a liquid the forces of cohesion do not hold so closely. The molecules can slide about over one another but they are still held by cohesive forces and prevented from escaping altogether. Only when a molecule gathers to itself sufficient kinetic energy (energy of movement) to overcome the force of cohesion does it fly away completely—in other words, vapourize.

According to the Dixon cohesion theory, the water in the vessels forms an unbroken column from the roots up to the topmost leaves. The cohesive forces in this column of water resist breaking, so that when the osmotic gradient set up by transpiration exerts a pull on the top end of this column of water, the whole column moves upwards just as the oil dipstick of a motor car comes up when you pull the top end of it. Notice that we have now got rid of the idea of suction and the difficulty of the water barometer. The force that pulls the water up is an osmotic one, and it is transmitted down the stem by the cohesion between the water molecules.

One is entitled to ask why we do not see this cohesive force operating in daily life? If you pour water into a narrow tube so that it is completely filled, why does not the water stay there when the tube is inverted, instead of pouring out? This would seem to be an immediate refutation of the Dixon theory. Yet the answer is simple. No man-made tube is *perfectly* clean; there are always some lurking particles of dust here and there. The Dixon theory assumes that the water wets the inside of the tube completely and so adheres to its walls. A few particles of dust would break this adhesion and supply the weak link in the chain that would allow the water to come tumbling out of the tube. This has been tested by experiment, using a rather neat little device known, for no obvious reason, as the 'water hammer'. This consists of a glass tube bent over in the form of a J. After scrupulously cleaning the tube it is almost filled with water, and then inverted. If the experiment is performed carefully enough the water remains in the tube and does not fall out (Fig. 19). The Dixon theory is vindicated.

The cohesive force that knits the water molecules seems to be considerable, and is ample for the job of holding together the water column in the vessels of a tree. Some measurements were made of the cohesive forces needed for the breaking open of a fern sporangium, and they gave a figure exceeding two hundred atmospheres for the strain that would finally break a water film. We have reason to believe that this figure was excessive, but it does seem that a tension

15.
Photomicrograph of
the embryo sac in a
flower bud of the lily

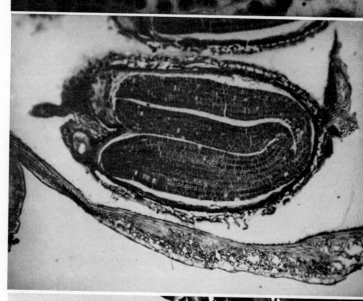

16.
Section through the
fruit of shepherd's
purse. The section
passes through a seed
nd the embryo that it
contains, showing the
o cotyledons and the
mbryo root or radicle

17.
Stem of the banana,
showing the way in
ch the fruit develops.
otograph by courtesy
lant Protection Ltd.)

18.
The honeysuckle, a
moth-pollinated flow
Note the pale colour
and long, tubular co

19.
Developing fruits o
the willow herb

of thirty atmospheres or so could be attained in a water column before it broke. This is sufficient for our needs.

It is one thing to arrive at a theory that will explain the upward movement of water in a plant, and quite another to prove that it is the correct one. It is here that the Dixon cohesion theory runs into difficulties. Nobody has succeeded in proving that it operates in Nature, and there is even some evidence that suggests that it cannot be the true explanation of the rise of sap. In one experiment, a pair of

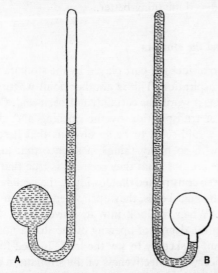

Fig. 19. The water hammer. The bulb **A** contains water. With care, the apparatus can be tilted so that the water runs into the tube, and then inverted, as shown in **B**, without disturbing the water, which is held by cohesive forces between its molecules

saw cuts were made on opposite sides of a tree trunk, one above the other, both cuts passing beyond the centre of the trunk, so that they overlapped one another. Theoretically this should have stopped the movement of water up the trunk, for all the columns of water in the vessels would have been broken by one cut or the other; yet it still rose.

We are not yet sure how to explain, or explain away, this difficulty. It is hard to believe that the water could have bypassed both cuts by taking a diagonal path up the stem, though it might have done. It may be that an unbroken column of water is not really necessary for the operation of the Dixon theory. Alternatively, the water may have travelled through the submicroscopic pores in the walls of the vessels instead of through their cavities. We shall not know until further experiments are carried out.

The present position of the Dixon cohesion theory of the ascent of sap may be said to be one of conditional acceptance. There are many difficulties in the way of accepting it unreservedly, but nobody to date has produced anything better.

Transpiration and the stomata

We have yet to consider the part played by the stomata of the leaf in 'controlling' transpiration. This is another point where opinions are divided and the last word has certainly not been said. On the face of it, the control of transpiration by the opening and closing of the stomatal pores would seem to be so obvious that it scarcely needs proof, but, as with so many things in science that are apparently 'obvious', things are not what they seem. It is true that in a general way stomata are so constructed that, owing to the peculiar thickenings of the walls of the guard cells, they open when turgid and close when water is scarce. When we look into it, however, we find that the relation between the degree of opening of the stomata and the water needs of the plant is sketchy to say the least. It used formerly to be the fashion to decry the effectiveness of the stomata in exerting anything but the barest minimum of control over transpiration, but latterly our ideas have undergone some revision, and it may be that the part played by the stomata in transpiration control may be greater than we had thought.

There are two things that will affect the rate of evaporation of water vapour from a leaf. One is the rate of passage through the stomata, and the other is the concentration of water vapour immediately outside the stomatal opening.

With regard to the first of these two things, we have already seen that owing to the formation of shells of diffusion (page 75) water vapour can diffuse through a series of small openings in a barrier nearly as fast as if the barrier were not there at all. The stomata are very numerous, and it might be expected that if their openings became smaller it might have little or no effect on the rate at which water vapour passed through them. In other words, if the stomata were to close somewhat, one would not necessarily expect it to have much effect on the rate of transpiration.

This has been found to be the case. Transpiration is not cut down immediately the stomata begin to close. How much they must close before producing an appreciable effect is a matter of conjecture, and varies very much according to the type of leaf, the structure of the stomata themselves, and the conditions prevailing at the time—particularly, whether it is a windy or a still day. Estimates vary from half open to as little as one fiftieth open before any real effect is produced.

Conditions outside the leaf, and especially the amount of wind, are very important in considering the effect of the stomata on transpiration. If the air is still, water vapour tends to collect over the mouth of the stoma, for there is nothing to blow it away. In such circumstances water vapour in the leaf cannot get away, even if the stomata are wide open. It is probable, therefore, that in really still conditions the degree of opening of the stomata is unimportant, and that they exert practically no control at all over transpiration.

On the other hand, if it is windy, water vapour cannot linger at the stomatal openings, for it is blown away as soon as it begins to collect. In such circumstances there is real evidence that the stomata are able to control transpiration with a considerable degree of accuracy.

When we come to the question of what makes the stomata open and close we again find that things are not quite what we might expect. The closing of stomata is due to loss of turgor in the guard cells; when the pressure of water in them falls below that in the neighbouring epidermal cells they close, opening again when their turgor rises to a value above that of the surrounding cells. One would expect, therefore, that when a leaf is wilting the stomata would close because they can no longer keep up the turgor in their guard cells. This is, in fact, roughly what they do, though there are various complications that we need not go into here.

Stomata and photosynthesis

If this were all that one could say about stomatal action the story would be a simple one, but it is not. Superimposed on the simple turgor mechanism there is a very complex relationship between the stomata and light. Plants vary very much in their stomatal behaviour but in general the stomata open in the light and close in the dark.

Thus there is a definite diurnal rhythm in the opening and closing of the stomata which is maintained irrespective of the water supply, often maintained even when the plant is kept under uniform conditions of illumination for a time.

This behaviour of the stomata is perhaps understandable if we realize that they are not only concerned with water loss. Stomata have an important function in photosynthesis, for it is through them that the precious carbon dioxide enters the leaf. If the stomata close, carbon dioxide cannot get in, and photosynthesis must stop. This, if it is maintained for a long period, as might happen during a prolonged drought, means starvation for the plant. The loss in weight of crops that results from a severe drought is as much due to starvation from closed stomata as to actual lack of water.

Seen thus, the behaviour of stomata with regard to light is entirely explicable, and this applies to most functions of plant elements. During daylight photosynthesis can go on, and the supply of carbon dioxide to the leaf has to be maintained: the stomata open if it is at all possible. At night, when it is dark, photosynthesis ceases for lack of light. The stomata might as well close to conserve water, and most of them do.

The nature of the mechanism that secures the opening of stomata in the light and their closing in the dark is far from clear, though most workers are satisfied that it is bound up with the acidity of the cell sap in the guard cells. A fall in acidity in some way produces opening of the stomata, while a rise in acidity encourages them to close. The change in acidity is probably due to variation in the amount of carbon dioxide that is present. When photosynthesis is active (as in broad daylight) carbon dioxide is used up, so that the acidity in the guard cells falls. At night, when photosynthesis stops, carbon dioxide is given off in the course of respiration, so that the acidity rises.

It is thought that the effect of change of acidity on the guard cells is bound up with the equilibrium between starch and sugar. When the cell sap becomes less acid, conditions are favourable for starch to be broken down to sugar by the enzyme phosphorylase. Starch is insoluble in water and so has no osmotic effect, while sugar, being soluble, is osmotically active. The appearance of sugar in place of starch in the guard cells raises the osmotic pressure of the cell sap, so that water passes in from the surrounding cells and the stomata, becoming turgid, open. Increase in the acidity, on the other hand, favours the conversion of sugar into starch, so that the reverse sequence of events takes place.

Many botanists are unsatisfied with this explanation of the effect of light on stomatal opening, and with reason. For one thing, the stomata of such plants as the onion, which never contain starch in their guard cells, still open in the light and close in the dark. There are various other objections to the theory. At the moment all that we can say of it is that it explains more facts than any other, but it is certain that further work will cause us to modify it, even if we do not abandon it in favour of something else.

Of what value is it to the plant to transpire such an enormous volume of water when only a minute percentage of it is actually used? Unfortunately our answer to that question is not as yet complete. A certain amount of water is of course needed in photosynthesis, and the passage of water through the plant keeps the solutes moving in the phloem (see Chapter 5), as well as bringing mineral salts up to the leaves. Lastly, on a hot day the evaporation of water from the leaves has a cooling effect which may be valuable; leaves in a greenhouse may scorch on a sunny day if the plants are prevented from transpiring by excessive humidity of the atmosphere. But none of these things can account for the huge extent of the transpiration stream. It is rather like going to Niagara to fill a bathtub. Plants seem to waste water with spendthrift abandon—water that is often very hard to come by.

The necessity for photosynthesis to go on whenever possible is the overriding factor here. For this to happen the leaves must be well ventilated to allow carbon dioxide to reach the photosynthesizing cells. The atmosphere contains no more than 0.03% of carbon dioxide, which is little enough when you consider that the bulk of a redwood tree is all fabricated from the gas. This immediately imposes a problem. If carbon dioxide can get into the leaf, water can just as certainly get out, for there is no possible way of securing one-way traffic through the stomata. The plant has to put up with the monumental loss of water, or it would starve. It is as simple as that.

Mineral intake of plants

In this chapter I have considered only the entry of water into the plant, and its passage through the plant and out into the air again. The roots of plants do not only absorb water from the soil, however;

they absorb mineral salts as well. All the many minerals that the plant needs to maintain life must enter the plant through its roots, for there is no other way in.

Let me say at once that we do not know how mineral salts are absorbed. To the schoolboy starting his first year of biology it is simple: salts enter the roots 'by osmosis'. This they certainly do not: such a phenomenon would be quite impossible. Osmosis may account for the entry of water, but not of solutes: the very fact that osmosis requires the presence of a semi-permeable membrane would see to that.

At one time it used to be thought that salts simply diffused into the roots; it was supposed that the cell membranes were permeable to some substances but not to others. The idea is pleasantly simple, but it cannot be true. For one thing, the cell sap of a plant is not just a concentration of the elements in the soil solution outside. More important, cells are able to accumulate certain substances against a concentration gradient, which could not happen if simple diffusion were at work. Certain seaweeds, for instance, accumulate relatively enormous quantities of iodine in their cells, and yet the concentration in the water outside is very small. Something more positive than just diffusion must be at work.

Modern research suggests that mineral intake by plants is bound up with the vital processes of living cells, and particularly with respiration. It has been shown that the absorption of minerals by slices of potato tuber increases as the rate of respiration rises. It seems likely that the passage of minerals across the cell membranes, and particularly across the membrane lining the vacuole (tonoplast), which seems to be the important one in mineral uptake, is connected with the electrical charges on the atoms (ions) that are being absorbed. Just how the process works, we have not yet been able to find out.

4 · How a plant feeds itself

The leaf of a plant is not only a manufacturing unit: it is the power supply for most of the world. Coal is nothing but the remains of past vegetation laid down in Carboniferous times, two hundred and seventy million years ago. Oil is of uncertain origin, but it is thought to be derived, at least in part, from plants. When we burn coal, we are burning the carbonized residues of past forests, and petrol in the tank of a motor car was once seaweed. Most of our electricity is derived from coal or oil power, both products of photosynthesis by plants. The only power that does not come from plants is found in the windmill and the water-wheel, plus a little that is derived from nuclear energy. Whenever we travel by road, rail or aeroplane we are propelled by the work of plants which long ago trapped the energy in sunlight and stored it up for future use.

Even if we walk we owe the energy that keeps our muscles moving to the activity of plants, for without plants there would be no food. The vegetarian owes his life directly to plants, for he eats nothing else (if he has the courage of his convictions). Those of us who prefer a more usual diet are equally in debt, for the flesh that we eat was nourished on plants at first, second or even third hand.

We do not know how long the miracle of photosynthesis has been in operation, but it must be as old as the plant kingdom. The first plants had chlorophyll, or something like it, and were able to feed themselves by taking the carbon dioxide out of the air and building it into sugar with the aid of sunlight, chlorophyll acting as an energy trap to steal the energy from the light and make it do chemical work. We can say with confidence, therefore, that photosynthesis has been going on for at least two thousand million years, and probably longer than that.

In most plants known to us today the leaf is the main seat of photosynthesis, though the stem may share in it, or even take over the

87

whole photosynthetic duty; but it was not always so. The first land plants had no leaves. One of our most ancient fossil land plants is *Zosterophyllum*, found in Silurian beds in Australia, which are about three hundred and fifty million years old. *Zosterophyllum* was a small plant, consisting of a tangled mass of branches which were probably underground, forming a rhizome system like that of mint or Solomon's seal. From this underground system rose erect branches which were evidently aerial, for they were covered with an epidermis and a cuticle, clearly designed to prevent the plant from drying up when exposed to the air. These branches were about a twelfth of an inch across, and were leafless, so that it must be presumed that they carried out the work of photosynthesis.

Leafless though the first land plants may have been, the leaf was not long in making its appearance and in taking over the main work of photosynthesis. In the same Australian beds that yield *Zosterophyllum* are found the remains of a plant called *Baragwanathia*. This was a larger plant than *Zosterophyllum*, with stems up to nearly three inches in diameter. These stems bore numerous closely-set leaves which were probably arranged in a spiral up the stem. The leaves were long and slender, measuring a fiftieth to a twenty-fifth of an inch in breadth and up to an inch and three-quarters in length. Each leaf had a single vascular strand ('vein') running down its centre and connecting with the vascular cylinder of the stem, clearly intended to bring water to the leaf and to carry the products of photosynthesis away.

Since the fossils of *Baragwanathia* are found in the same Silurian beds as *Zosterophyllum* the two plants must have been of comparable age. The evolution of the leaf cannot therefore have been long delayed once plants had established themselves on dry land.

As the principal organ of photosynthesis, the leaf is beautifully constructed to serve its purpose. It is usually thin and flat, thus exposing a large area of surface not only for the intake of carbon dioxide but also for the sun to illuminate. It is true that some leaves are not like this, but these have been modified in response to some other factor, notably the need to conserve water. The interior of a leaf is well ventilated, so that carbon dioxide can reach the photosynthesizing cells without delay. In addition, a leaf has in its veins abundant vascular strands which not only bring water to the leaf, but also carry away in their phloem the organic compounds formed by photosynthesis. Lastly, a leaf is well covered with epidermis and cuticle, these preventing undue evaporation of water, and yet has in its stomata means of ready access of carbon dioxide to the internal air spaces.

Response of plants to gravity

Plant stems usually grow away from the force of gravity and towards the light, thus carrying their leaves up where the light is strongest. If a seed is sown upside down, the young root does not rise above the soil and the stem does not burrow downwards. The effect of gravity makes itself felt as soon as the young stem and root emerge from the seed coat, the stem turning upwards while the root bends over and grows down. This response to gravity is built into the plant; it does not have to 'learn', any more than a new-born piglet has to learn where to find its mother's teats.

The response to gravity comes about because of a substance called auxin that is produced by the growing tips of stems and roots. The effect of auxin on the growing tip of the stem is to make the newly formed cells elongate faster, so that the growth of the stem is speeded up. Normally the auxin passes backwards from the tip of the stem, producing its effects equally all round. If a stem is placed on its side, however, the auxin concentration on the lower side becomes greater than on the upper—we are not certain how, though we suspect it is a question of differing electrical potentials. The result is that the lower side of the stem, with its larger concentration of auxin, grows faster, so that the stem turns upwards. When it is growing vertically again, equilibrium is restored.

The response of a root to gravity is brought about in the same way, though the root reacts differently. Auxin appears to slow down the growth of roots; consequently, when a root is laid on its side the increase in concentration of auxin on the lower side slows down the growth relative to the upper side, so that the root curves downwards.

Botanists were much intrigued at the apparent difference in reaction of roots and stems to the same substance, auxin. Tests showed that root auxin is identical with stem auxin; it is the response that is different. The matter was cleared up when it was shown that *very small* quantities of auxin actually stimulated the growth of roots instead of retarding it; similarly, large doses of auxin retarded the growth of stems. A general principle covering both root and stem was thus established: small doses of auxin stimulate growth, while large doses retard it. The difference between a root and a stem lies in the interpretation of the term 'a small dose of auxin'. Roots appear to be more sensitive than stems in their reactions to auxin; what is a small dose for a stem is a large dose for a root.

When and how stems and roots acquired this difference in reaction we do not know, but we can guess that it came soon after the functional difference between stems and roots was established. Auxin is a very old substance. It is found in the eggs of certain seaweeds and in various other parts of the algae (plants of the seaweed group). Presumably, when plants first left the sea they took with them the capacity of making auxin and reacting to it, and turned it to good account in their new life ashore. This is another example of the remarkable economy that we see in so many of the things that plants do. Sometimes they appear to have slipped—as in the prodigious waste of pollen that goes on in wind pollination—but in the main they live and die with the greatest possible economy of effort.

Growth towards light

Besides growing away from gravity, a plant stem grows towards the light. This reaction can be seen by anyone who grows a potted plant near a window, for the stem bends in the direction from which the light is coming. This reaction is again brought about by auxin, which concentrates more on the dark than on the illuminated side of a stem; the darkened side grows faster, causing the stem to bend towards the light. The growth of stems towards the light, like their growth away from gravity, helps to bring the leaves up where they will receive the greatest amount of light and air to enable them to carry out their work of photosynthesis.

With certain exceptions, such as the climbing roots of the ivy and the underground roots of the white mustard, roots are unaffected by light in their direction of growth.

When a plant is grown in poor lighting conditions, for instance if it is surrounded and shaded by other plants, certain changes take place in the growth of the stem. It becomes long and spindly, with its leaves separated by abnormally long intervals, and is then said to be 'etiolated'.

Another result of etiolation is a pale green colour, for chlorophyll is not developed in the absence of light. The full effects of etiolation are best seen on a plant kept entirely in the dark. The stem becomes long and weak, and the whole plant is of a yellowish colour, quite unlike the healthy green of a normal plant. The leaves also are much smaller and more widely separated, and have the same unhealthy yellowish colour as the stem. It is worthy of note that growth in length of the stem is actually *faster* in the dark than in the light, for light tends to slow down growth.

The phenomena associated with etiolation are of value to a plant that finds itself being overshadowed by its neighbours. By growing up quickly it may eventually reach the light and have a chance to make normal growth thereafter. Provided that etiolation has not gone too far, a plant soon recovers when it is exposed to full illumination.

The main ingredients for photosynthesis are, as we have seen, carbon dioxide and water—two of the simplest compounds known to chemistry. Also necessary are chlorophyll, and light to activate it. Certain other essentials are less obvious. For instance, there must be present in the leaf the sugar ribulose, together with groups of atoms known as phosphate groups—so-called because when combined with hydrogen, they form phosphoric acid.

Chlorophyll and other pigments

The chlorophyll in a leaf is not distributed evenly throughout the protoplasm, but is contained in minute lozenge-shaped bodies called 'chloroplasts'. Examination of chloroplasts by means of the electron microscope, an instrument that can give us far greater magnifications than the optical microscope, has confirmed the opinion already held: that they have a very complex structure.

The chlorophyll in a chloroplast is contained in submicroscopic particles called 'grana', which are embedded in a colourless body called the 'stroma.' A single granum does not consist entirely of chlorophyll, but is built up like a multi-decker sandwich. The chlorophyll molecules are in layers, one molecule thick, between layers of protein and 'lipid' (fatty substance). The layers are arranged like this: protein, chlorophyll, lipid, chlorophyll, protein, chlorophyll, lipid and so on, right through the granum, in regular order. A cell of the palisade of the leaf will contain many chloroplasts. Each chloroplast will contain many grana, and each granum has this complex structure of layers.

The function of the chlorophyll is to trap the energy in light and hand it on in a form suitable for chemical work. When chlorophyll is brightly illuminated it becomes excited—in the chemical sense, not the colloquial. The electrons that spin round its atoms jump into higher orbits, and in so doing absorb energy. Later on they sink back to their original orbits, and when they do so the energy that they absorbed is given up. In this lies the secret of the energy-trapping power of the chlorophyll molecule.

The chemical structure of chlorophyll is complicated. There are, in fact, several different forms of chlorophyll, known as chlorophyll *a*, chlorophyll *b*, chlorophyll *c* and so on; of these, chlorophyll *a* and *b* are found in the chloroplasts of higher plants, the other being confined to the various groups of algae (the seaweeds, and what the Irishman called the 'fresh-water seaweeds'). Chlorophyll *a* is blue-green in colour, while chlorophyll *b* is yellow-green. They are contained in the chloroplasts in the proportions of about three molecules of chlorophyll *a* to one of chlorophyll *b*, the combination giving the grass-green colour of most higher plants. It appears to be chlorophyll *a* that is mainly responsible for photosynthesis, though it is assisted by chlorophyll *b* in a way that will be explained in a moment.

Besides the two chlorophylls, the chloroplasts contain two other pigments—or, to be precise, two other series of pigments. These are the 'xanthophylls' and the 'carotenes'. They vary in colour from yellow to red. Like the chlorophylls, xanthophyll and carotene are photochemically active. Although they do not play a direct part in photosynthesis they are of importance indirectly.

White light is really a mixture of all the colours of the rainbow, as can be shown if the light is allowed to pass through a glass prism, when it is separated into the various colours of the spectrum. If we pass a beam of white light through a solution of chlorophyll in a suitable solvent, such as ether, and then examine it through a spectroscope (an instrument containing a prism to separate white light into its various colours), we find that there are differences in the extent to which the different colours are transmitted by the chlorophyll. Green light is transmitted strongly, but much of the red, and also much of the blue-violet, is absorbed by the chlorophyll and not transmitted at all. In other words, the spectrum of chlorophyll shows dark areas called absorption bands in the red part of the spectrum, also in the blue violet part.

If the energy of light is to be used in photosynthesis the light must first be absorbed by the chlorophyll. Since red light is strongly absorbed by chlorophyll, we might expect that photosynthesis would be active in red light, and this is found to be the case. If we cover a potted plant with a glass bell jar coloured red and measure the amount of photosynthesis that takes place with a given amount of illumination for a given time, we find that it is somewhat greater than the amount of photosynthesis in a plant in blue light, and very much more than the photosynthesis in green light.

If chlorophyll *a* were the only pigment present in leaves, the plant would only be able to use light of a narrow range of wavelength for photosynthesis; light of all other colours would be wasted as far as sugar-making was concerned. The process would be highly inefficient, for much of the white light striking the leaves would be useless. The presence of chlorophyll *b*, and of the xanthophyll and carotene, improves matters. Their absorption spectra are different from that of chlorophyll *a*: they can absorb light from different parts of the spectrum, and in so doing their molecules become excited. It is true that these substances do not appear to play any direct part in photosynthesis itself, but the excitation of their molecules can be passed on to chlorophyll *a*, which can then use the energy it has gained photosynthetically.

This seems to be the function of the xanthophyll and carotene in the chloroplasts, and probably of the chlorophyll *b* also. By absorbing light of wavelengths that chlorophyll *a* cannot deal with, and then passing on the energy gained to chlorophyll *a*, they make photosynthesis possible over a broader range of wavelengths than would otherwise be available.

Even with the help of the accessory pigments, as the xanthophyll and carotene are called, photosynthesis does not take place equally in light of all colours. There is still a bad spot in the green, for instance. But white light is more effective in the presence of accessory pigments than it would be without them.

Since photosynthesis depends on the absorption of light, we may ask ourselves how efficient the leaves of plants are in absorbing the light that falls upon them. Our evidence points to a high degree of efficiency. A single chloroplast may absorb from thirty to sixty per cent of the light that actually falls on it. Since the chloroplasts are fairly densely arranged in the palisade tissue, we may fairly assume that most of the light that falls on a leaf will strike a chloroplast somewhere. Even the light that fails to do so, or light that is reflected from the chloroplast without being absorbed, is not necessarily lost to photosynthesis. There are plenty of air spaces in the leaf which, in conjunction with the walls of the mesophyll cells, make reflecting surfaces that may turn the light back before it finally passes through the leaf. This reflected light is still available for photosynthesis should it be snapped up by a chloroplast.

Different kinds of leaves vary in the amount of light they reflect—one has only to look at a few plants to see that. It may even be an advantage to the plant to have leaves that reflect much of the light instead of absorbing it, for over-bright illumination can be harmful to chlorophyll. In general, one finds that plants habitually living in strong sunlight have smaller and thicker leaves than those in shade. Some plants, notably the beech, produce 'sun leaves' near their tops, where the solar radiation is strongest, and 'shade leaves' lower down, where there is less light. In the beech the sun leaves are thicker than the shade leaves, with a two-layered palisade tissue.

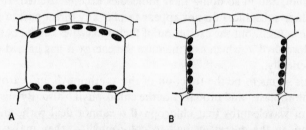

Fig. 20. A cell of the leaf of the duckweed, showing the arrangement of the chloroplasts, A, in weak light and B, in strong light

Plants can often vary the amount of light absorbed by their leaves according to the circumstances. The individual chloroplasts of a leaf may change their shape in response to the amount of illumination they are receiving, becoming rounder in poor light, flatter in bright light, and at the same time will orient themselves with their edges towards the light, so as to present as small a surface as possible. The chloroplasts may also change their position according to the intensity of the light, spreading out over the illuminated sides of the cells in poor light, while in strong light they concentrate themselves along the vertical, less illuminated cell walls (Fig. 20). This can be seen very well in the leaves of the ivy duckweed (*Lemna trisulca*).

In many shade plants, such as the moschatel (*Adoxa moschatellina*) and the wood sorrel (*Oxalis acetosella*) the cells of the epidermis are notably convex on their outer sides. This causes them to act as convex lenses, focusing the light into the interior of the cell where the chloroplasts are gathered.

The leaves of trees are commonly arranged in a mosaic, each leaf being set so that, as far as possible, it does not shade the leaves immediately beneath it. In this way the fullest use is made of available light. The beech is an example of a good leaf mosaic. This is why the beech casts such a dense shade which, with the aid of the shallow-feeding roots of the beech, keeps the floor of a beechwood practically clear of other plants. You have only to look upwards in a beechwood on a summer day to see the effect of the leaf mosaic.

Many plants turn their leaves on their leaf-stalks so as to present the flat of the leaf to moderate light, while the edge of the leaf is turned towards a strong light. The French bean (*Phaseolus vulgaris*) is a particularly good example of this.

The carbon dioxide present in the air enters the leaf through the stomata and passes into the intercellular spaces. The walls of the mesophyll cells are wet, and the carbon dioxide dissolves in the water, forming carbonic acid. In this form it reaches the chloroplasts.

It must be remembered that air contains but 0·03 per cent of carbon dioxide—three parts in ten thousand. In spite of this very low concentration of carbon dioxide, plants manage to get all they need. The leaf must be considered to be a very efficient organ for straining carbon dioxide out of the air.

The other raw material needed for photosynthesis is water. This comes from the soil, and is carried from the roots up the stem and to the leaves by the xylem vessels or tracheids, finally reaching the cells of the mesophyll through the veins of the leaf. There the water, the carbon dioxide and the chlorophyll come together, and if the sun is shining the result is sugar.

The chemistry of photosynthesis

The chemistry of photosynthesis is very complicated, and I can do no more than hint at it here. Basically, it may be represented by the chemical equation

$$CO_2 + H_2O = (CH_2O) + O_2$$

This indicates that carbon dioxide has combined with water to form the compound (CH_2O), which represents an unspecified carbohydrate, and that oxygen gas is given off as a byproduct. This equation is merely a conventional way of expressing the reaction, and means very little. For one thing, the 'carbohydrate' CH_2O does not exist; the formula CH_2O is just a convenient way of representing the carbohydrate that is formed as a result of photosynthesis, be it glucose ($C_6H_{12}O_6$), cane sugar ($C_{12}H_{22}O_{11}$) or something else. For another, the equation suggests that the whole process is carried out in one step, whereas in fact it proceeds through a series of stages.

We have seen that the process of photosynthesis needs the energy from light to bring it about, but that does not mean that light is necessary for every stage. There has been much controversy in the past about the stage at which the light energy is actually used—the so-called 'light' reaction, as opposed to the 'dark' reactions which do not need the intervention of light. It now appears that the light reaction is probably the photolysis of water: that is, the splitting up of water with the aid of light energy. This can be represented by the equation

$$H_2O = H + OH$$

but this again is an understatement. The hydrogen atoms are presumed to combine with some substance, forming compounds called 'hydrogen donors' because they readily give up the hydrogen atoms they have acquired to some other substance, being regenerated in the process so that they can pick up more hydrogen atoms formed by the photolysis of water. One substance known to be present in plant cells that might behave in this way is nicotinamide adenine dinucleotide phosphate—known as NADP for short.

The OH part of the water molecule must also be disposed of. This can be represented thus:

$$4OH = 2H_2O + O_2$$

though, like the other reactions that we have discussed, it is more complex then the equation seems to suggest. It may be that the OH groups of atoms first form hydrogen peroxide, thus:

$$4OH = 2H_2O_2$$

which then splits up into water and oxygen

$$4H_2O_2 = 2H_2O + O_2$$

However, it is more probable that some organic peroxide is formed and then decomposed.

One thing we are certain about. All the oxygen given off in the

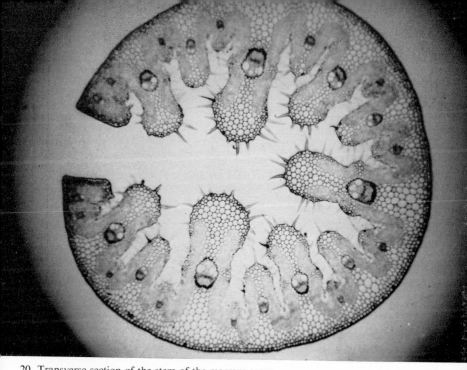

20. Transverse section of the stem of the marram grass

21. Transverse section of part of the photosynthetic cortex of a twig of *Casuarina*. Note the palisade-like cells packed with chloroplasts

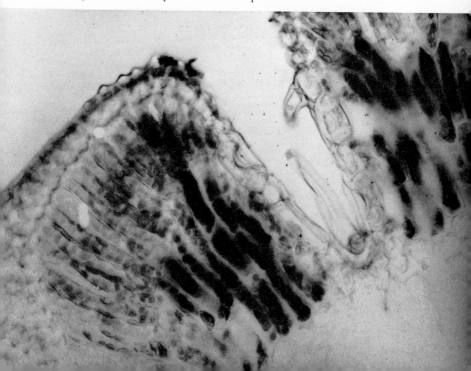

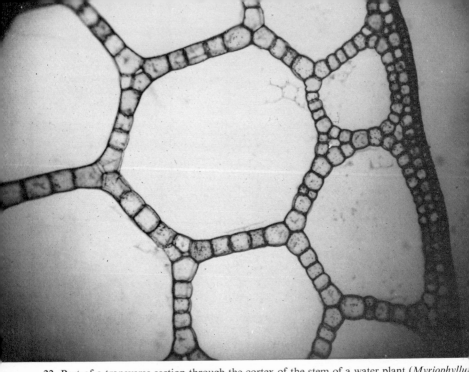

22. Part of a transverse section through the cortex of the stem of a water plant (*Myriophyllu*
showing the large air spaces

23. Transverse section of the leaf of the pine

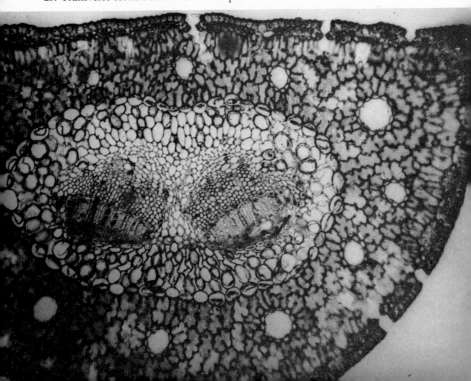

course of photosynthesis, as shown in the first equation I quoted, comes from the water and not from the carbon dioxide.

Looking again at the equation

$$CO_2 + H_2O = (CH_2O) + O_2$$

we see that the hypothetical carbohydrate (CH_2O) contains both less oxygen and more hydrogen than carbon dioxide (CO_2). This means that, in photosynthesis, carbon dioxide has been reduced. This is a 'dark' reaction: light is not needed for it to take place. Before it can be reduced the carbon dioxide must be 'fixed'—that is, made to combine with some substance in the leaf. After much research and argument, we are pretty certain that we now know what this substance is: it appears to be ribulose diphosphate.

The substance ribulose (see earlier mention on page 91) is itself a carbohydrate, with the chemical formula $C_5H_{10}O_5$; because its molecule contains five carbon atoms it is known as a pentose sugar. When the ribulose molecule combines with carbon dioxide it momentarily contains six carbon atoms, the number contained in the glucose molecule, $C_6H_{12}O_6$. One might think that once the carbon dioxide had combined with ribulose and been reduced that would be the end of the matter, for glucose is a common sugar in Nature and would serve well as the end-product of photosynthesis. It is not as simple as that. When the carbon dioxide combines with the ribulose the six-carbon molecule so formed breaks down into two three-carbon molecules of phosphoglyceric acid.

This may seem to be an unnecessary complication, for we have produced a molecule containing six carbon atoms only to break it down again. There is more to it that that: we have not so far observed any pointless or totally wasteful plant activity, and this is not one either. Phosphoglyceric acid is an intermediate product in the process of respiration, in which carbohydrate is oxidized to carbon dioxide and water. Respiration is the opposite process to photosynthesis, and in the formation of phosphoglyceric acid the two processes meet, with a whole host of exciting biochemical possibilities. With phospho-glyceric acid as a starting point all sorts of things can be synthesized in a comparatively simple way.

We must remember also that ribulose diphosphate has somehow to be regenerated somewhere during photosynthesis. If it were not, photosynthesis would cease as soon as supplies of ribulose diphosphate ran out. Experimental evidence suggests that ribulose diphosphate may be reformed from phosphoglyceric acid as fast as it is used up.

The phosphoglyceric acid formed in photosynthesis appears to be converted into phosphoglyceraldehyde. This reaction involves 'reduction' of the phosphoglyceric acid (any chemical action in which oxygen is removed from a molecule, or hydrogen added, is called a reduction process). This is where the hydrogen formed in the photolysis of water comes in. You will remember that this was supposed to have combined with NADP to form reduced NADP, or NADPH. The reduced NADP gives up its hydrogen to the phosphoglyceric acid (in chemical terms, it acts as a hydrogen donor), and the NADP thus regenerated can then combine with more hydrogen from water, forming NADPH again. The process is thus cyclic and can go on, theoretically, for ever.

Once we have arrived at phosphoglyceraldehyde an almost infinite array of possibilities opens up before us. From phosphoglyceraldehyde as a starting point it is theoretically possible to synthesize almost anything, and there is little doubt that most, if not all, the complex organic compounds that go to make up a plant trace their origins, by more or less devious means, from this compound. Nitrates taken in with the soil water, after reduction, are combined with carbon compounds to form amino-acids, the organic acids that are bricks out of which proteins are built. Similarly, phosphates provide the phosphatic groups used in building complex nucleotides and nucleic acids, including the universal DNA (deoxyribonucleic acid) that now appears to be the basis of all life. That is why plants are so self-sufficient: given a supply of carbon dioxide and a few simple chemicals of the kind one finds in bottles in any laboratory, they can manufacture for themselves all they need.

In this very brief account of photosynthesis I have inevitably oversimplified the process. This book is not about plant physiology, except inasmuch as it contributes to the design of plants. Readers who wish to know more about it will find plenty of specialist books on the subject.

As a result of photosynthesis, sugar appears in the leaves of plants during daylight. In most plants this sugar is not allowed to accumulate. Some may be carried away from the leaves at once by the phloem of the vascular bundles in the veins, a process called translocation (see Chapter 5). Only a relatively small amount can be carried away at a time, and the bulk of the sugar is stored, until nightfall puts a stop to its accumulation by photosynthesis. In the meantime, the sugar is converted into starch. This is a relatively easy process, involving the withdrawal of water from molecules of glucose which in the process of losing water become strung together into the large 'super-molecules' of starch. In this way the sugar is converted into an insoluble substance which because of its insolubility will not upset the osmotic arrangements of the leaf cells.

The starch that accumulates in the daytime, known as 'transient starch', is at night converted into sugar again and removed from the leaves. This process will be the subject of the next chapter.

5 · The transportation system in plants

We saw in the last chapter that the leaves of a plant are the centres of photosynthesis, where sugar is manufactured. In a simple seaweed there is no serious transportation problem. The sugar is, in the main, used where it is made; only in the larger seaweeds is any appreciable amount transported from one part of the plant to another. When plants increased in size and complexity transportation of materials became an urgent problem. In a tree, both mineral salts and manufactured food materials may have to be moved from one end of the tree to the other—a distance, in a redwood or a eucalyptus, of two or three hundred feet. This calls for a quite sophisticated system of transport.

Transportation of organic compounds

Water we have already dealt with in Chapter 3. Mineral salts absorbed through the roots travel, in the main, with the water in the vessels and tracheids. The movement of organic compounds, such as sugar that has been elaborated in the leaves, is a different story. This movement of materials from one part of the plant to another is known as translocation. It was recognized soon after botany became a science that the translocation of organic compounds in plants was a separate problem from the movement of water. Water and minerals travel in the xylem of the vascular bundles, but it is otherwise with the sugars and other organic compounds.

On certain occasions the water in the wood of trees may contain sugar—sometimes a great deal of sugar. The sugar maple (*Acer saccharum*) is an example. In late winter, water withdrawn from the xylem may contain a high concentration of sugar, the sugar disappearing in spring when the leaves unfold. This is because the living parenchyma cells of the wood store up large quantities of starch. In the main, the wood cannot be the chief transporting system for organic compounds. For one thing, the transportation stream in the wood is upwards, towards the leaves, whereas in summer the leaves are sending sugar downwards, towards the roots. Dissolved organic compounds could hardly travel against the stream.

It is now known that the translocation of organic compounds takes place through the phloem of the vascular bundles (see Chapter 2). The cells particularly concerned are the 'sieve tubes', with their companion cells, that make up the bulk of the phloem in most higher plants.

A sieve tube, when mature, is a somewhat elongated cell, the end walls of which are perforated by a series of fine holes—hence its

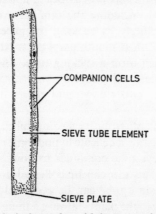

COMPANION CELLS

SIEVE TUBE ELEMENT

SIEVE PLATE

Fig. 21. A sieve tube, with its companion cell

name. Adjoining sieve tubes are placed end to end, communicating by their sieve plates, as the perforated end walls are called (Fig. 21). A sieve tube element has a thin lining of protoplasm, but no nucleus— a very unusual arrangement. Lying alongside the sieve tube is its companion cell, narrower than the sieve tube and filled with protoplasm. The companion cell contains a nucleus which apparently has to do double duty, controlling the activities of the sieve tube to be as well as the compound cell.

The fact that organic compounds are transported by the phloem has been proved by many experiments. If the bark, including the phloem, is removed in a complete ring from the stem of a woody plant—an operation known as 'ringing'—it is easy to show that organic substances, including both carbohydrate and nitrogen compounds, accumulate above the ring while the parts of the plant below the ring are starved. Removal of the phloem evidently prevents the transport downwards of these substances. Then there is the evidence derived from some classical experiments on cotton plants. It was shown that an increase in the sugar and organic nitrogen content of the leaves was followed by a similar increase of these substances in the bark; when the sugar and nitrogen level fell in the leaves there was a corresponding fall in the bark. In the wood, on the other hand, there was no such variation.

In recent years some very elegant techniques have been perfected for finding out just what the sieve tubes contain in a living plant. One of these uses the sap-sucking proclivities of aphides—small insects of the order Hemiptera, including the familiar greenfly and blackfly. These insects live entirely by sucking the juices of plants, and to enable them to do so their mouth parts are modified into a series of very fine stylets which form a sucking tube. The stylets are pushed into the leaf of a plant and waggled about until they enter a sieve tube. The insect then settles down and gorges itself on the juice of the plant.

If the aphis is cut off from its feeding tubes so as to leave them undisturbed, they provide a delicate hypodermic needle through which the sap from the sieve tube continues to flow, often for as long as twenty-four hours. This sap can be collected and analysed, so that the contents of the sieve tube can be accurately determined with a minimum of damage to the plant—always a thing to be avoided where possible, for the behaviour of a plant being tormented in a laboratory may be very different from that which happens under natural conditions.

Experiments such as this have put the translocation of organic substances by the phloem beyond doubt. Inorganic substances may also be translocated in the phloem, for it has been shown that while the upward transportation of minerals takes place mainly in the wood, the downward translocation of these occurs in the phloem (when it occurs at all). Some minerals, such as calcium, do not seem to be capable of downward translocation; once they have reached the leaves they are there for good.

Numerous experiments have shown that the chief carbohydrate carried in the phloem is sucrose (cane sugar). Flowering plants appear to specialize in this sugar for phloem conduction, in contrast to the algae (seaweeds) in which there is no phloem to conduct and sucrose is not so easily formed. Why sucrose should be the particular carbohydrate chosen we do not know, but it has a relatively high molecular weight and therefore low osmotic activity compared with a monosaccharide such as glucose, and is also easily broken up into glucose and fructose when these sugars are needed.

Translocation by the phloem is a phenomenon which can be exploited in the application of certain hormone weedkillers such as 2, 4-D. This substance, which is not unlike auxin, is carried in the phloem with the downward carbohydrate stream. If it is sprayed on to the leaves of a plant it will be carried in the phloem to any underground part such as roots or rhizomes where sugar is stored, and so kill these as well as the foliage. This is very useful in dealing with such weeds as bindweed, which cannot be eradicated in any other way because any portion of rhizome left alive deep in the soil will sprout to give a new plant.

Nitrogen is carried in the phloem in the form of amino-acids, of which asparagine is particularly important. Amino-acids are organic acids containing nitrogen. When strung together by what are called peptide linkages they form more complex organic compounds. The simplest is glycine (amino-acetic acid), $CH_2(NH_2) \cdot COOH$. The group of atoms-COOH is acidic, while the group-NH_2 is basic. Two molecules of glycine can therefore combine with one another to form diglycine, like this:

$$2 \ CH_2(NH_2) \cdot COOH = CH_2(NH_2) \cdot CONHCH_2 \cdot COOH + H_2O.$$

You will notice that diglycine is itself an amino-acid, for its molecule contains both the acidic group-COOH and the amino group-NH_2. It can therefore combine with another amino-acid by a peptide linkage. This sort of thing can go on indefinitely; moreover, different amino-acids can combine with one another in this way, so that vastly complex molecules containing thousand of amino-acid molecules may be formed. This is how the proteins, the complex nitrogenous compounds that are the basis of all life, are built up.

Among the organic compounds carried by the phloem are certain substances that control growth and reproduction, much as hormones act in the animal kingdom. These substances are in fact known as 'plant hormones', and an important example is auxin. As described in the last chapter, auxin controls the growth of stems and roots. We shall discuss other examples of plant hormones later.

Use of foods

We have not yet considered where all the food carried by the phloem is going. Food needs to be translocated away from the leaves for three general purposes. It is needed by growing organs, such as the tips of the stem and root, for building new tissues. It is also needed by all the living cells in the plant for respiration—the process by which all living things, animal and plant, get their energy. Finally, food in excess of the needs of the moment is stored, usually in some fairly insoluble form, until it is needed.

Let us consider for a moment a germinating seed. The seed contains in its 'seed leaves' or cotyledons, or in its endosperm (see Chapter 6), a great deal of stored-up food in the form of oil or starch, together with a greater or less amount of protein. These foods are insoluble, and so the first thing that must happen when the seed germinates is the breaking-down of insoluble food materials to a soluble form, so that they can be translocated by the phloem from the place of storage to the place where they are going to be used—in this case the growing region. The breakdown of the stored food is brought about by a class of substances called 'enzymes', which are organic catalysts, able to promote a chemical action without becoming permanently changed themselves.

Different enzymes break down the various classes of food material. Diastase breaks down starch into malt sugar (maltose), lipase breaks down fats into glycerine and fatty acids, while proteases break down proteins into their constituent amino-acids. The products of breakdown are then transported to the growing regions where respiration is most active and where new tissues are being formed.

In the ripening of a fruit the reverse takes place. The young fruit contains rudimentary seeds in the process of formation. Here the soluble substances, transported by the phloem from the actively photosynthesizing leaves, are converted into insoluble form for storage.

The approach of autumn is always a busy time for storage organs, equalled in activity only by the onset of spring. Photosynthesis is slowing down, and in the leaves some mineral matter is being 'returned to store'. The fronds of the bracken fern, for instance, are rich in potash during the summer, but in the winter much of the potash has returned to the underground rhizome, awaiting another spring.

Not all the mineral matter is conserved. I have already mentioned calcium as an element that stays in the leaves, once it has reached them. This is useful because in the course of their chemical activities plants produce a certain amount of oxalic acid, a potent poison. In an animal this would no doubt be excreted along with the rest of the waste matter of the body, but in plants excretion in the animal sense hardly occurs. Plants have no kidneys. Instead, the harmful oxalic acid is made to combine with calcium, forming calcium oxalate. In some of the cells of many leaves crystals of calcium oxalate can be seen. This, being an insoluble compound, is harmless; though present in the cells, its insolubility prevents it from entering into plant metabolism. In much the same way, mercuric chloride (corrosive sublimate) is a deadly poison to animals, while mercurous chloride (calomel) is far less so. Mercuric chloride is soluble in water, while calomel is virtually insoluble. When the leaves of trees are shed in the winter the crystals of calcium oxalate go with them. In this way the dangerous oxalic acid is finally removed from the plant.

It must not be thought that plants never excrete anything, for some of them, at any rate, excrete quite actively through their roots. Sometimes the excretions of plant roots are put to practical use by parasites waiting to attack them, as in the sinister story of the potato root eelworm and the excretions of the potato plant that give it the signal to attack.

Plants are sometimes said to 'excrete' carbon dioxide through their roots, evidence being quoted of roots etching a pattern on blocks of marble with which they are grown in contact. This is not excretion in the usual sense of the term, however, for carbon dioxide is a normal product of respiration.

The actual mechanism by which the translocation of food materials in the phloem comes about is at present imperfectly understood. The mass flow theory, put forward in 1930 by Munch, is generally accepted, but only because no better hypothesis has yet been found. According to Munch, the phloem is to be regarded as a continuous osmotic system which extends all through the plant. In parts of the plant where organic compounds are being synthesized, such as the leaves, the products of synthesis are fed into the system by an active biological process that requires the energy of respiration to keep it going. This means that there will be an increase in osmotic pressure in the parts of the phloem where loading is going on. In other parts of the plant the phloem is unloading its store of organic compounds to cells that need them; here there will be a decrease in osmotic pressure. Water is taken in where the osmotic pressure is high (that is, where loading is going on) and given out where the osmotic pressure is low (where the phloem is unloading). As a result of the taking in and giving out of water, there is a mass flow of water through the system, from loading points to unloading points. The water of course comes from and returns to the xylem, which always accompanies the phloem.

There are plenty of snags attached to Munch's theory, but so far nobody has been able to disprove it. If it could be shown that a substance, or two different substances, could move in opposite directions in the same sieve tube, the theory would immediately be rendered untenable, but so far this has not been demonstrated.

6 · The problem of reproduction

The higher plants reproduce sexually by means of seeds—or, to put it more accurately, a seed is the result of the sexual act. By higher plants here I mean seed plants—the flowering plants and their more humble allies, the conifers. Lower plants—the ferns and their kind, the mosses and liverworts, the algae and the fungi—have their own methods of reproduction in which seeds are not formed. It is with the flowering plants that this chapter is mainly concerned.

Components of flowers

Flowers vary enormously in their shape, their colour, their size and their degree of complexity, but—fortunately, perhaps, for the systematic botanist—they are all variations on the same theme. Basically, a flower consists of four series of parts, attached to the tip of the flower stalk, usually somewhat swollen, which is called the receptacle.

First, we have the sepals, forming the outer part of the flower. The sepals are usually green, and before the flower opens they enclose it and protect it from harm. Sometimes, however, the sepals are petaloid—that is, coloured like petals, from which it may be difficult or impossible to distinguish them. In some flowers the sepals are joined together, in others quite separate. All the sepals, taken together, constitute the calyx (Fig. 22).

Second, we have the petals. These are usually but not always showy and brightly coloured, and are the principal organs that, by their brightness, attract pollinating insects. Like the sepals, the petals may be joined together, or they may be free from one another. The petals are known collectively as the corolla.

Within the corolla we find the stamens. These are the male elements in the flower, and they produce the pollen grains (microspores) with which the flower is pollinated. A stamen usually consists of a pair of anthers, which contain the pollen, attached to a stalk called the fila-ment. The stamens are together known as the androecium.

Finally, in the centre of the flower is the ovary or gynoecium. This is the female part of the flower. The gynoecium consists of one or (usually) more carpels. Each carpel is in three parts. The main part is a hollow ovary, above which is a structure called the stigma, attached to the carpel by a long or short stalk called the style. Usually the carpels forming the gynoecium are fused together forming a compound structure, from the top of which one or several styles arise, but in some flowers, for example buttercups and magnolias, they are separate.

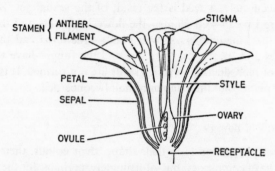

Fig. 22. Diagram of a flower, cut in half
to show the arrangement of the floral parts

Notice that the terms 'ovary' and 'gynoecium' are not quite synonymous, though in practice they are often used interchangeably. Strictly speaking, the ovary comprises the lower, hollow part of the carpel or carpels, while the term gynoecium refers to the whole structure. The gynoecium of a flower was formerly called the pistil, and the old name still lingers on.

Inside the ovary one can see from one to many tiny, round white objects. These are the ovules. They will later become the seeds, but before they can do so the flower must be pollinated, and the ovules fertilized.

Various other accessory parts may or may not be present in a flower. Important among these are nectaries, patches of glandular tissue that secrete a sugary liquid, nectar, which is attractive to bees and other insects, and so helps to ensure pollination. A flower that is without nectaries may still be insect pollinated, for nectaries are not the only things that attract insects to flowers. Bees, for instance, are great collectors of pollen, and a flower that has no nectar but is visited by bees for the sake of its pollen is called a pollen flower.

Often, just below a flower, there is a small leaf known as a bract.

A typical flower consists of calyx, corolla, androecium and gynoecium, but all the parts may not be present. Quite commonly either the calyx or the corolla is absent, in which case it is usual to call whichever is present the perianth. The term perianth is also used to include calyx and corolla where both are present.

A flower that has both male parts (androecium) and female parts (gynoecium) is said to be 'hermaphrodite'. Most flowers are like this, but some are unisexual—they have either androecium or gynoecium, but not both. Such flowers may properly be spoken of as male or female. Sometimes both male and female flowers are found in the same plant, in which case the plant is said to be 'monoecious'. In 'dioecious' plants the whole plant is either male or female.

I mentioned in Chapter 2 that the flowering plants can be divided into two groups, the dicotyledons and the monocotyledons, according to whether the number of 'seed leaves' is two or one. This division is fundamental, and shows itself in various ways. Dicotyledons have leaves with the veins arranged in a network, a single ring of vascular bundles in their stem, and usually show secondary growth by means of a cambium (Chapter 2). Monocotyledons usually have strap-like leaves with parallel veins, vascular bundles scattered throughout their stems, and, except in rare instances, no secondary growth.

The division between dicotyledons and monocotyledons also shows up in the flowers. You can usually recognize the flower of a dicotyledon because each set of floral parts—sepals, petals, stamens and carpels—is four or five in number, or a multiple of four or five. In the flowers of monocotyledons, on the other hand, the parts are in threes or multiples of three.

Of course there are plenty of exceptions to these general rules— botany would not be half the fun it is if all plants conformed to a few strict rules. I sometimes tease elementary students by showing them a plant of herb Paris (*Paris quadrifolia*) and asking them whether it is a dicotyledon or a monocotyledon. This plant has broad leaves with the veins arranged in a network, and its flowers have their parts arranged in fours, but it is a perfectly good monocotyledon, belonging to the family Trilliaceae, allied to the lily family. Things are not always what they seem.

Pollination and fertilization

Before a flower can set seed it must be fertilized, and before fertilization it must be pollinated; pollination and fertilization are by no means the same thing. The act of pollination consists of the transfer of pollen, from the stamens of the same or another flower, to the stigma. It would not be untrue to say that the flower as a structure has evolved in order to ensure that pollination takes place—usually with pollen coming from another flower, bringing about cross-pollination. Some flowers, for example the wheat are, generally pollinated with their own pollen (self-pollination), but this is exceptional.

The anther of a stamen contains four cavities or pollen sacs in which the pollen grains lie packed together. When mature the anther opens by two longitudinal slits, exposing the pollen. Normally, especially in insect-pollinated flowers, the pollen remains until it is dislodged by an insect brushing against it; in wind-pollinated flowers it is dislodged by the anther shaking in the wind. Sometimes, especially in wind-pollinated flowers, the pollen is shaken out of the anthers by a special mechanism. A well-known example of such mechanisms is that of the pellitory-of-the-wall (*Parietaria diffusa*). This plant, found in cracks in old walls and in hedgebanks, has separate male and female flowers. In the male flower the filaments are elastic, and before the flower opens they are held arched inwards by the boat-shaped segments of the perianth. As the flower opens the anthers are freed with a jerk, and as they fly back the pollen is discharged in a little cloud.

Pollen grains vary very much from species to species; so much, indeed, that it is often possible to tell what plant a particular sample of pollen has come from by looking at the pollen under a microscope. In this way much information has been gathered about the distribution of trees in the past, for beds of peat yield pollen grains that can still be identified after many thousands of years. Pollen grains are usually round or oval in outline, though there are some exceptions. The pollen grains of the eel-grasses (*Zostera*) are thread-like, coiling round the stigmas of the female flowers. *Zostera* is a genus of marine flowering plants, growing in the sea just below low tide mark, and it is thought that this peculiar shape may be an advantage in enabling the pollen to float.

The size of pollen grains is fairly constant for a given species, but in different species it varies very much. The pollen grains of the forget-me-not are about 0·0001 inch in diameter, while those of the cucumber are a hundred times bigger.

The surface of a pollen grain may be smooth or rough, wind-pollinated plants tending to have smooth pollen while that of insect-pollinated plants is more usually rough; the rough coat helps the pollen to stick to the insects' legs, as well as causing the pollen grains to stick together. Sometimes, as in certain of the Onagraceae (evening primrose family), the pollen grains have several threads of a sticky substance attached to them which tend to bind them together.

Pollen transferred to the stigma of a flower sticks there, often assisted by the roughness of the receptive surface, which is frequently sticky as well. The viscid secretions not only help the pollen grains to stick, but probably promote their germination.

Pollen grains vary in the ease with which they will germinate. In some species, they germinate very readily, either in sugar solutions or, in some cases, in water. The strength of the solution that pollen grains will tolerate varies very much; pollen of the apple will germinate in sugar solutions of from 2·5 to ten per cent concentration, that of the snowdrop does better in a concentration of from one to three per cent, while a species of senna has been found in which the pollen will germinate in seventy per cent sugar solution. The pollen of the prim-rose, among other plants, germinates in distilled water but not in tap water. In some cases the pollen will not germinate at all except in the presence of substances secreted by the stigma. The pollen of a species of *Mussaenda*, a Burmese plant, was found to germinate only when a piece of the stigma was included in the water in which the grains were placed, though the stigmatic tissue could be replaced by the sugar fructose, and nothing else. Many cases are also known where pollen will only germinate on the stigma of a flower of the same species, or a closely related species.

The advantage of having a definite range of concentration of sugar or some other substance that is necessary for the germination of pollen is not far to seek. It guards against the premature germination of the pollen before it reaches the stigma, thus avoiding waste of pollen. Yet the number of species with pollen that will germinate in water seems to indicate that the advantage is not great.

When a pollen grain germinates a tubular structure, called the pollen tube, grows out of it and penetrates the tissue of the stigma (Fig. 23). The means by which the solid matter of the stigma is penetrated vary with different species. In some the pollen tube grows through the actual cells of the stigmatic tissue, dissolving them away by means of enzymes in the same way as one of the threads or hyphae of a parasitic fungus such as *Pythium* grows through plant tissues. In other cases the cells of the stigma are loose and offer little resistance to the pollen tube, which pushes its way between them. In still other instances there may be definite canals in the stigma through which the pollen tubes grow. The objective of the pollen tube is, of course, the ovary.

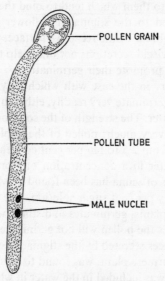

POLLEN GRAIN

POLLEN TUBE

MALE NUCLEI

Fig. 23. A germinating pollen grain,
with its pollen tube

The pollen tube must have something to guide it during its journey which, in a flower with a long stigma, may be very long indeed for such a tiny thing. In most cases the guiding force seems to be twofold: the tube grows away from air and towards certain chemicals. If pollen grains are persuaded to germinate on the surface of gelatine, their tubes immediately turn into the gelatine, away from the outside air. If germinating pollen grains are presented with pieces of stigma, they grow towards them. In most cases the chemical that attracts the pollen tube has turned out to be a sugar, but not always; in *Narcissus tazetta* and various other plants the active substance has been shown to be a protein.

112

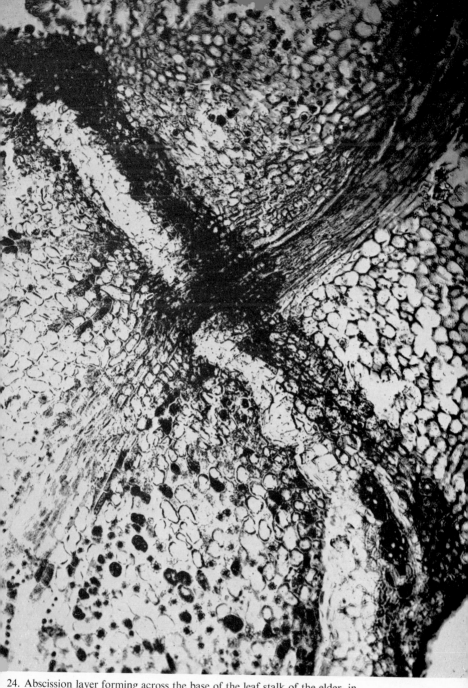

24. Abscission layer forming across the base of the leaf stalk of the elder, in preparation for leaf-fall

25.
Climbing stem of the
runner bean twining
round a stake. The
direction of twining is
anti-clockwise

26.
Tendril of the sweet
grasping a stake

When it reaches the cavity of the ovary the pollen tube grows towards an ovule, which it enters usually by a small hole called the 'micropyle'. The pollen tube contains two nuclei near its tip; these are the two male sex nuclei or gametes that will effect the fertilization of the ovule. The ovule develops as a little mound of tissue which becomes the nucellus of the ovule—the tissue that is later enclosed in an outer covering or integument. The integument develops later as a ring-like upgrowth from the base of the ovule; there may be one integument, or two. The integument or integuments grow up round the ovule, covering the nucellus completely save for a spot at the end of the ovule which the integument just fails to enclose, so that a hole is left, exposing the surface of the nucellus. This is the micropyle, through which the pollen tube enters the ovule.

During the early growth of the ovule the end where the micropyle is to develop points away from the placenta, the part of the ovary where the base of the ovule is situated. In some plants the ovule remains in this position, but in most it twists round on itself during its development, so that the micropyle finally comes to lie next to the stalk of the ovule, facing the placenta. While all this is going on, there develops in the nucellus of the ovule, at the micropylar end, a cell called the 'megaspore mother cell'. This divides twice, forming a row of four cells, perpendicular to the surface of the ovule. Three of these subsequently degenerate, but the fourth—usually the innermost one—becomes known as the 'functional megaspore', at a later stage called the 'embryo sac'. This cell becomes much larger, absorbing most of the nucellus as it does so. At first it contains only a single nucleus, but later it divides, and the two daughter nuclei so formed move to opposite ends of the cell. Then each divides twice, forming four nuclei at each end of the embryo sac. Of the four nuclei at the end farthest from the micropyle, three become separated by cell walls and are then called the 'antipodal cells', while the fourth wanders towards the centre of the embryo sac. Of the four nuclei at the micropylar end of the ovule, three become surrounded by protoplasm and are then known as the 'egg apparatus', consisting of an egg and two 'synergidae' or 'help cells'. The remaining nucleus wanders towards the centre of the embryo sac where it meets the nucleus from the other end. These two nuclei are called the 'polar nuclei'; eventually they fuse to form the 'primary endosperm nucleus' (Fig. 24).

Details of the happenings in the embryo sac vary in different plants. The two polar nuclei, for instance, may postpone their fusion until

113

later. These differences in detail do not affect the general picture.

Asexual (spore) reproduction

In the fern you will remember that there are two generations in the life history—the sporophyte that bears the spores, and the gametophyte that bears the sex organs. The sporophyte, which is the fern plant, reproduces asexually by means of spores, which germinate to produce the prothallus, a tiny plant no more than a quarter of an inch across. The gametophyte bears the sex organs, and the product of its

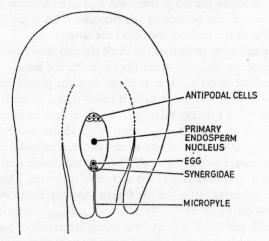

ANTIPODAL CELLS

PRIMARY
ENDOSPERM
NUCLEUS

EGG

SYNERGIDAE

MICROPYLE

Fig. 24. Diagram of a section of an ovule,
showing the embryo sac

sexual reproduction is the fern plant. This has already been described on page 28.

In some of the relatives of the fern, such as the small club moss (*Selaginella*), there are two kinds of spore and two kinds of gametophyte. The smaller spores, called microspores, germinate to produce the male gametophyte, which in this case is so much reduced that it consists only of a few cells. The larger spores, called megaspores, produce on germination a small mass of green tissue on which the female sex organs are developed. The sperms produced by the male gametophyte fertilize the eggs on the female gametophyte, and so a new plant of *Selaginella* is produced.

In the flowering plant the gametophyte generation has been still more reduced, although it can still be recognized if one looks for it. Like *Selaginella*, the flowering plant has two kinds of spores. The pollen grains are the microspores, but the male gametophyte is so reduced that it really does not exist; we merely have the two male gametes at the end of the pollen tube, and another nucleus called the 'tube nucleus'. The female gametophyte is represented by the embryo sac in the ovule. In neither the male nor the female gametophytes can we find anything that can even remotely be described as a plant.

The advantage gained by this extreme reduction of the gametophyte generation is that the plant is now completely independent of water for its sexual reproduction. There is no delicate gametophytic plant requiring a damp place in which to live, and the elimination of the swimming sperm has made the emancipation from water complete. A flowering plant can reproduce sexually in the middle of a desert, thanks to that masterpiece of design, the seed.

Reproduction in gymnosperms

The other great group of seed plants, about which I have as yet said little, are the gymnosperms; they include such common coniferous trees as the pines, the fir, the spruce fir, the larch and the cedar, as well as a great many interesting but less familiar plants. The gymnosperms came into existence long before the flowering plants; the first ones were already well established in late Devonian times, nearly three hundred million years ago, and more than one hundred million years before the flowering plants appeared on the scene.

In the gymnosperms there are no flowers. Instead, they produce male and female cones, consisting of scales spirally arranged around a central axis and bearing pollen sacs or ovules according to the sex of the cone. The most important difference botanically between the gymnosperms and the flowering plants is that gymnosperm ovules are not contained in carpels, but lie exposed on the scale of the cone. The word gymnosperm means 'naked seed'. When a gymnosperm is pollinated the pollen grain must reach the ovule directly, entering through the micropyle and producing its pollen tube afterwards.

The gymnosperms are less advanced than the flowering plants. They have the advantage of producing seed, with all that goes with it in the way of not being dependent on water for reproduction, but their reproductive process is not quite so beautifully specialized as it is in the flowering plants. Their ovules, not being enclosed in carpels, are less well protected than the ovules of flowering plants, and this fact of their not having carpels means they do not have fruits, with all the opportunities for dispersal that go with them. The gymnosperms are earlier models than the flowering plants, and their design still has imperfections that have since largely been put right. In particular, they began the course of their long evolution before the insects had really got going, so that they still depend on wind for pollination.

All the same, it must not be thought that the gymnosperms are not a successful group. In the Douglas fir and the redwood they have two of the world's largest trees, and their distribution is world-wide. In high northern latitudes they dominate all other vegetation, and there is no sign whatever that they are in any way declining.

Development of seed

After an ovule has been fertilized, it develops into a seed. The act of fertilization consists of the fusion of one of the two male nuclei from the pollen tube with the egg nucleus. In the gymnosperms this is the only nuclear fusion that takes place, for the second male nucleus has no work to do, and usually disintegrates soon after its fellow has fertilized the ovule. In the flowering plants, however, there is a double fertilization. One of the male nuclei fuses with the egg nucleus, while the other one unites with the primary endosperm nucleus. In the further development of the ovule into a seed this fusion nucleus divides again and again, forming a mass of tissue called the 'endosperm'. This is packed with food material, and serves to nourish the embryo plant during its early development.

After fertilization, the ovule enlarges as it becomes a seed. From the fertilized egg develops the embryo, consisting, in the case of a dicotyledon, of a pair of 'seed leaves' or cotyledons, with the 'plumule' or young stem between them. In monocotyledons only a single cotyledon is developed, and the plumule appears at one side. In both dicotyledons and monocotyledons the young root or 'radicle' appears at the opposite end of the embryo. At the base of the cotyledons there is also formed a structure called the 'suspensor'. This is composed of a number of flat cells, and by its elongation it pushes the developing embryo towards the centre of the seed, where it is surrounded and nourished by the endosperm.

As the embryo develops it absorbs some of the endosperm. The amount that is absorbed varies in different plants. In many dicotyledons the whole of the endosperm is absorbed, the cotyledons becoming thick and fleshy as a result. Thus the swollen cotyledons make up the bulk of the mature seed, and when the seed germinates they pass on their stored food to the growing seedling, enabling it to survive the period that must elapse before its leaves have developed sufficiently to feed it by photosynthesis.

In other dicotyledons, and most monocotyledons, this does not happen. The cotyledons remain thin, and the bulk of the reserve food remains in the endosperm until germination, when it is used up by the growing seedling. The sunflower is an example of a dicotyledon that behaves like this, and maize and wheat are monocotyledons that also retain a large amount of stored food in their endosperms until germination.

While all this is going on inside the seed, the integument is developing into the hard outer skin, or 'testa', with which seeds are surrounded. Finally the seed is mature, and passes into a dormant condition. In the dormant state everything is in a state of 'suspended animation'. Growth has stopped and even respiration, the universal need for all living cells, has slowed down to a point where it is no longer detectable. The seed remains dormant until germination awakens it into life again.

The length of the resting period varies a great deal. In some tropical mangroves there is no dormant period. The seed germinates while it is still borne by the parent plant, and when it is released from the fruit it plunges downwards like a dart, the point of its young root burying itself in the mud where the mangrove grows. At the other extreme, some seeds require a lengthy period of dormancy before they can be persuaded to germinate.

Germination

All seeds need moisture, a suitable temperature, and oxygen for respiration before they will germinate. Some seeds also need light—or darkness. The seeds of the mistletoe (*Viscum album*) must be exposed to light before they will germinate; if kept in the dark they will not do so. Certain varieties of lettuce behave in the same way. On the other hand, onion seeds should be kept in the dark, for their germination is slowed down, and may be prevented altogether, by exposure to light.

117

The effect of light on the germination of seeds has nothing to do with photosynthesis, for at that stage photosynthesis has not begun. We do not know how light produces its effect. One can only presume that some delicate physiological process going on in the seed is in some way upset if the seed gets too much light in some cases, or too little in others. Some seeds will germinate in darkness at low temperatures but become light-requiring if the temperature is raised. Lettuce seeds of a variety that does not need light for germination become light-requiring if they are treated with coumarin, a substance that is known to inhibit germination. This suggests that, in a light-requiring seed, the effect of light may be to break down some germination inhibitor that is present in the seed.

Where all the conditions necessary for germination are present, including light or the absence of it, and the seed still does not germinate, we know that the seed is dormant. It is still capable of germinating and will germinate when it is ready, but the time is not yet ripe. The embryo remains locked away in its seed coat waiting for something to happen that will release it from its prison to begin the process of growing into a new plant.

Sometimes dormancy may be due to purely physical causes, such as the extreme impermeability of the seed coat preventing the entry of water. This is so with certain sweet pea seeds. If the seed coat is scratched with a file before the seeds are sown they germinate more readily. Occasionally it is not water but oxygen that cannot penetrate the seed coat, as in *Xanthium*. Again, the seed coat may be so tough that until decay has softened it the imprisoned embryo is unable to expand. In Nature, the attacks of bacteria and fungi in the soil are usually sufficient to deal with these stubborn seed coats.

A tough seed coat is far from being the only cause of dormancy. In some plants, when the seed is otherwise ripe the embryo itself has not developed to a stage when it can germinate. Seeds of this kind will obviously require a period of further development before germination can begin; those of some members of the orchid family are particularly prone to this type of dormancy.

Some seeds containing a fully developed embryo at the time of ripening will still not germinate until the expiration of a period known as 'after-ripening' has elapsed. In the rose family, for instance, this process of after-ripening may take six months. This ensures that seeds produced in the autumn of one year will not germinate until the following spring—an advantage to the plant, for if a spell of mild conditions during winter should tempt the seeds to germinate prematurely, the seedlings might well be killed by a following frost.

The details of the after-ripening process are not properly understood. Sometimes the seed contains a growth-inhibiting substance which must be removed before germination can begin. In some desert plants rain is needed to wash the germination inhibitor out of the seed. Such seeds will only germinate after heavy rain; a light shower is not sufficient to remove the inhibitor. This again is clearly an advantage to the plant, for if the seed were to germinate after the first shower the seedlings would be dried up when the rain ceased. Only heavy rain, sufficient to provide for the subsequent growth of the plant, will induce germination.

Other seeds appear to need the development of growth-promoting substances before they will germinate. The formation of these substances is induced by various external factors, particularly the exposure to prolonged low temperatures under moist conditions. Apple seeds are of this kind. Here again, the conditions needed before germination are just what the seeds may expect during a normal winter season.

Viability of seeds

The longevity of seeds is a subject that has been obscured by many exaggerated reports in the past. *All* seeds eventually lose the power to germinate if they are kept long enough, even under ideal conditions, but the seeds of different plants vary very much in their ability to withstand storage. Some seeds will only germinate when fresh from the pod. The seeds of the wood sorrel (*Oxalis acetosella*) are unable to withstand drying, and those of cycads also appear to be short-lived. On the other hand, the seeds of some plants have been known to germinate after storage for many years. Those of the dock (*Rumex crispus*) have been known to germinate after a period of eighty years. The longest period of dormancy known with certainty is that of some seeds of the Indian lotus (*Nelumbo nucifera*) that were kept in the herbarium of the British Museum and which were soaked with water when the Museum was bombed in 1940. They germinated, in spite of the fact that the date on the herbarium sheet showed that they were two hundred and thirty-seven years old.

Seeds of the Indian lotus have also been recovered from peat beds in Manchuria and successfully germinated. These may have been as much as four hundred years old, but the age of the peat is not known with certainty. Stories of Indian lotus seeds germinating when four thousand years old, though they gained much notoriety at the time of publication, are almost certainly false, as also is the widely held belief that grains of wheat contained in ancient Egyptian tombs have been known to germinate. The seed of the wheat plant is rather short-lived, as seeds go, and could not possibly last even a hundredth of this age and retain the power of germination.

Recently, a report has appeared of some seeds of the Indian lotus germinating after being buried in burrows made by the lemming, ten thousand years ago. If the age of these seeds is substantiated, we may have to change our ideas about the potential longevity of seeds.

The seed is not the only means of reproduction available to plants. Every gardener knows that many plants are propagated by means of bulbs, corms, tubers, runners, cuttings and other forms of accessory reproduction where the setting of seed is not required. The term 'vegetative reproduction' is used to cover the multitude of ways in which plants are able to spread without the intervention of the sexual process that results in the formation of seed.

Vegetative reproduction

The strawberry runner is a well-known example of an effective system of vegetative reproduction. Towards the end of the flowering season the strawberry plant produces long, prostrate stems (runners) that creep over the surface of the ground, bearing a few small, scale-like leaves. Eventually the tip of the runner turns upwards and produces a new plant, the runner being continued by a new branch that arises in the axil of a leaf—that is, the angle formed between the leaf and the stem. Where the new plant forms a fresh rosette of leaves, roots develop from the stem (adventitious roots) and bury themselves in the soil. After the new plant has become established, the part of the runner between the parent part and the new one dies away, leaving the new plant to live an independent life.

A feature of the strawberry runner is the speed at which it grows. This carries the new plants well away from the parent, serving to reduce competition. An extension of the strawberry type of runner is seen in the blackberry, which has long stems that rise into the air and then droop downwards. When the tip of a stem touches the ground it swells and then takes root, establishing a new bramble bush. In a few years the new plant may become independent. The formation of new plants in this way accounts for the impenetrability of bramble thickets.

Sometimes a root may develop an adventitious shoot that grows up above ground and forms a new plant. That is why the creeping thistle (*Cirsium arvense*) is so hard to eradicate. The main tap root produces many whitish lateral roots which creep underground to a distance before they form their adventitious shoots.

A very common organ of vegetative reproduction is the rhizome. This is an underground stem which serves the plant in much the same way as the runner of the strawberry, except that it is at all times protected from frost by the soil. The rhizomes of couch grass (*Agropyron repens*) will be familiar to anyone who has had the job of exterminating it in the garden. Mint and lily-of-the-valley are also common rhizomatous plants. Solomon's seal is named from the round scars—the so-called 'seals'—left on the underground rhizomes by aerial shoots that have died away. The rhizome of Solomon's seal is thick and fleshy because it is used as an underground food store where nourishment can be laid up for the winter, serving to feed the young plants that grow up from it in the spring.

The specific names 'repens', 'reptans' or 'radicans' applied to a plant often refer to the presence of runners or rhizomes. The creeping buttercup (*Ranunculus repens*), and the much rarer species, *R. reptans*, found on lake margins in the Lake District and in Scotland, are examples.

The association between food storage and vegetative reproduction is more strongly seen in the stem tuber, of which the potato is an example. The potato plant forms, just below ground level, a number of lateral branches which end in swollen structures, the tubers. Each has a protective covering of periderm (the 'peel'), and bears a number of buds in the axils of scale leaves (the 'eyes'). If a potato is cut into a number of pieces, all bearing an 'eye', and the pieces are planted in the soil, each will grow into a new plant. The true nature of the 'eyes' of the potato can be seen when tubers are put out in boxes to 'chit' or sprout before planting. Each 'eye' grows out into a shoot bearing embryo leaves.

It should be noted that the potato is not really a root crop, though it is often called one. Roots do not bear scale leaves and buds; the tubers are part of the stem.

Sometimes the root of a plant is tuberous, as in the spotted orchid (*Orchis fuchsii*) and other members of the genus *Orchis*. The tops of

121

the roots are swollen and contain stored food to give the spring growth a good start. As with stem tubers, more than one root tuber may be formed by a plant, so that the root tuber can act as a means of vegetative reproduction.

The corm of a crocus is formed by a swelling of the base of the stem. If you dig up a growing crocus plant during April you can see that a new corm is forming among the bases of the leaves, on top of the old one, while the remains of the corm of the year before can often be seen, much shrivelled, below the corm that has just produced its flowers. Later in the year the new corm will have grown to its full size. It will be shrouded with scale leaves, in the axils of which are the buds that will grow into new flowering shoots in the following year.

Corms are quite common in wild plants as well as cultivated species. The bulbous buttercup (*Ranunculus bulbosus*), one of the commonest of the buttercups, distinguished by its turned-down sepals, and the cuckoo pint (*Arum maculatum*), both show well-developed corms.

In a corm the reserve food is stored in the swollen base of the stem. In a bulb, such as that of the tulip or the daffodil, the food is stored in the thick, fleshy leaves, while the stem from which they spring is reduced to a flat, disc-like structure. In the tulip bulb there is a central bud surrounded by fleshy leaves. In the spring this bud gives rise to the flowering shoot with its leaves, using up the food stored in the fleshy leaves of the bulb. In the axil of one of the fleshy leaves is a bud which, fed by photosynthesis from the expanded foliage leaves, increases in size and forms the bulb of the following year. Sometimes more than one axillary bud may develop into a bulb, so that there will be two or more bulbs the following year; vegetative reproduction is thus achieved.

A different kind of bulb is seen in the daffodil (*Narcissus pseudonarcissus*) and the snowdrop (*Galanthus nivalis*). Here the fleshy bulb scales are formed from the bases of the leaves of the previous year's growth, or of several preceding years in the daffodil. The foliage leaves arise from the old stem, which persists from year to year, while the flower arises on an axillary branch. The main or terminal bud persists, with the stem, from year to year, and the bases of the foliage leaves swell at the end of the flowering season to form the new season's bulb. New bulbs are formed in the axils of the outer bulb scales.

The essential difference between the tulip and the daffodil bulbs is that in the tulip bulb a complete new structure is produced every year, whereas in the daffodil and the snowdrop part of the bulb, at any rate, is permanent. In botanical language, we say that the tulip bulb arises by 'sympodial' growth, while in the bulbs of the daffodil and the snowdrop the growth is 'monopodial'.

The bulb of the bluebell (*Endymion nonscriptus*) shows a mixture of tulip and daffodil characteristics. It is produced by sympodial growth, like the tulip, but resembles the daffodil in being formed mainly by swollen leaf bases.

Some plants reproduce vegetatively by means of 'bulbils', small buds that are easily detached from the plant and fall on to the soil to grow into new plants. Like bulbs, they usually have their leaves packed with reserve food. The lesser celandine (*Ranunculus ficaria*) is a common weed with bulbils. In the crow garlic (*Allium vineale* var. *compactum*) bulbils replace the flowers in the inflorescence.

To the gardener, vegetative reproduction is a convenient way of ensuring that the progeny will be similar to the parent. When a cutting is taken from a plant we know that the new plant will be exactly like the old in all important respects, since there has been no mixing of hereditary material from two parents. Many of our cultivated plants are propagated only in this way, for they are hybrids produced by the crossing of two different varieties, and so would not breed true from seed.

Anybody who sowed the seed of a Victoria plum in the hope of raising another Victoria plum tree would almost certainly be disappointed. The Victoria plum is the product of the crossing of many varieties of different plum trees, and the produce of its seed would almost certainly bear very little resemblance to the parent tree. It is true that one *might* raise a new and superior variety in this way, but the chances of this happening are infinitesimal. If we want to propagate a Victoria plum in all its luscious perfection, we must do it by grafting. Only in this way can we be sure that the offspring will resemble the parent in all respects.

It is the same with the potato. 'Seed' potatoes are not seeds at all: they are small tubers that have grown from the stem of another potato plant. The true seeds of the potato plant are formed from flowers in the normal way, but if they are sown they do not reproduce the variety from which they came, but something quite different, and probably useless as an economic plant. Occasionally a seedling may give rise to a new and valuable variety, but for every one that does so, tens of thousands have to be thrown away.

Advantages of seeding

Among wild plants vegetative reproduction is very common. It has certain advantages over sexual reproduction: particularly, it is simple and reliable, and helps the species to spread quickly over a limited area. Most of the seeds formed by wild plants are wasted for one reason or another. They may fall on stony ground, like the seed in the parable of the sower; they may be eaten by birds; they may be smothered by other plants before the seedlings can get established. Vegetative reproduction, on the other hand, is far more certain. Yet it has also its disadvantages. Although it is useful, as we have seen, in enabling a plant to spread quickly over a limited area, it cannot secure dispersal to a distance. The young plant must necessarily grow up fairly close to the parent. The seed has available all the ingenious methods that have evolved over the years for its dispersal, which may be measured in miles—sometimes hundreds of miles—instead of inches or feet.

The seed can also offer the advantages of dormancy to tide over the winter, when vegetative life above ground must come to a halt. A seed is the safest way in which an embryo plant can be packed to resist the worst that winter has to offer. Indeed, for an annual plant, it is the only way. A seed is safe not only against the cold of winter but also against other hardships. In the case of a desert plant, for instance, it is not the ice and snow of winter that have to be feared, but the blistering heat of the sun. In the long run, the real importance of the seed stems from the fact that it is a product of sexual reproduction, and therefore contains the possibility of variation. When the male gamete from the pollen grain fuses with the egg nucleus in the ovule there is a mingling of characters of male and female parents. This is so even when a flower is self-pollinated, and is much more so when cross-pollination takes place. The seedling is a little different from either of the parents. This gives natural selection the raw material on which to work, and it is only through variation that evolution can take place at all. If variety is the spice of life, of evolution it is the very essence.

7 · Pollination

Pollination consists, as I explained in the last chapter, of the transfer of pollen from the stamens to the stigma of the same or another flower. In most plants cross-pollination is the rule. Many plants have evolved ingenious mechanisms to secure it. The advantage of cross-pollination, as already mentioned, is that characters from two different parents are thereby blended, and if these parents are different plants there are possibilities that new characters may arise in the resultant plant.

Pollination can be effected through a number of agencies, the most important of which are wind and insects. It used to be thought that wind-pollinated flowers came first, insect pollination following afterwards by the adaptation of mechanisms originally intended for the agency of wind. Now we are not so sure. It must be remembered that insects came into the world well before the flowering plants; in Carboniferous times there were dragonflies with a wing span of twenty-seven inches busy over the swamps in the coal forests. The Carboniferous insects were primitive creatures as insects go, but later they began the evolutionary explosion that might well have made them masters of the world, were it not for certain ineradicable faults in their design. The evolution of the insects into all the twenty-eight orders that we have today took place at the same time as the evolution of flowering plants into the dominant plant form. What could be more natural than that the two groups should evolve in partnership, the insects feeding on the flowers and, in return for the nectar that the flowers soon began to produce, carrying the pollen from flower to flower? There is a considerable following today for the idea that insect pollination came first, and that wind pollination followed. Of course, the two methods of pollination may have evolved side by side. In the present state of our knowledge we just cannot tell.

The earliest flowers were probably something like the flower of the magnolia—an open kind of flower with an indefinite number of perianth segments, no distinction between calyx and corolla, many stamens producing a large amount of pollen, and a gynoecium in which the carpels were quite separate from one another. From such a flower it is easy to see the origin of the buttercup. Today flowers of this type are pollinated by a variety of different insects; there is no specialization of structure that limits the pollination to one particular kind of insect visitor, such as a bee or a moth.

The buttercup flower is a good example of this rather primitive type of floral structure. The name 'buttercup' is loosely applied to several different species, the three commonest being the bulbous

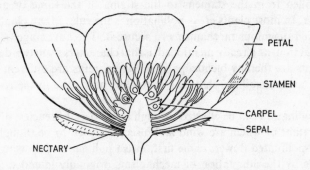

Fig. 25. The flower of a buttercup,
cut in half to show the floral parts

buttercup (*Ranunculus bulbosus*), the creeping buttercup (*R. repens*) and the meadow buttercup (*R. acris*). All have a similar type of flower. There are five sepals and five petals which are not joined together, and the whole flower is wide open to admit any kind of insect. There are many widely spreading stamens, and the gynoecium, in the centre of the flower, consists of many separate carpels on a raised part of the receptacle—the swollen end of the flower stalk to which the floral parts are attached. At the base of each petal, on the inside and hidden by a tiny scale, there is a nectary (Fig. 25).

Many different kinds of insect visit the buttercup flower, some for pollen, such as the tiny moth *Micropteryx calthella*, but most of them for nectar. Butterflies and moths, bees, wasps, beetles, bugs and flies are all found on flowers of the buttercup, which provides a landing stage for insects large and small. Thrips are also found there, sucking the sap from the petals.

Any of these insects may be responsible for carrying pollen from flower to flower, and with such a variety of visitors one would expect, in the absence of any special mechanism to prevent it, that self-pollination would often occur. In fact, it is rare. This is because the buttercups are markedly self sterile: that is, they are physiologically incapable of being fertilized by their own pollen. With just a few individuals this is possible, but for most buttercups it is an impossibility. Cross-pollination is thus assured, except in rare instances.

Self-sterility is by no means uncommon among flowering plants, and it may be due to a number of causes. Usually the pollen will not germinate on the stigma of the same flower, or, if germination occurs, the growth of the pollen tube soon stops. In the laburnum (*Laburnum anagyroides*) the pollen will not germinate unless the stigma has been slightly damaged; small wounds made by the bee that transfers the pollen are sufficient to induce germination. Self-sterility is characteristic of certain varieties of apple, such as Blenheim Orange, and many varieties of pear, as every fruit grower knows. Specimens of these varieties will not set fruit if grown by themselves.

Specialization for ensuring pollination

As flowers advanced in their evolution they became specialized for pollination by certain kinds of insects, the bee being the pollinator *par excellence*. The bee has a long tongue which is thrust into a flower in order to get at concealed nectar. Thus we find that the nectary in bee-pollinated flowers—and still more in flowers pollinated by butterflies and moths—is tucked away at the bottom of a tube formed by the petals, which are commonly fused together. Only an insect with a long tongue can get at it, and smaller insects are often prevented from crawling down the corolla tube and stealing the nectar by rows of hairs that bar their passage, though some of them have learned to steal the nectar by boring through the corolla tube near the nectary.

A great deal of work has been done during the present century on the senses of bees, and we are beginning to understand how they are drawn to certain kinds of flowers. Three things seem to be important: the colour of the flowers, their scent, and the pattern of markings on some of them that Christian Sprengel called 'Saftmale' in 1793, and which are today generally spoken of as 'honey guides'. For a long time it was thought that bees had no colour vision at all, and that they saw everything as a pattern of grey, the colours being represented merely as different shades of intensity. In 1914 Von Frisch began his classic researches into the senses of bees, and he soon proved that they have a good sense of colour, though different from our own. Von Frisch trained bees to take syrup from dishes placed on squares of blue. When the bees had become accustomed to associating a blue square with food, he next presented them with a whole series of squares, one of which was blue and all the others grey of varying intensities—some darker and some lighter than the blue square. None of the squares had any food associated with it. The bees were not bamboozled in the slightest degree: they went straight to the blue square and looked for the syrup. Clearly they can distinguish blue from grey without difficulty.

Repeating his experiment with different colours, Von Frisch was able to show that bees have a keen sense of colour, though it works differently from our own colour sense. A man can distinguish about sixty different shades of colour—an artist more—but a bee can only distinguish four colours. Starting from orange, and proceeding along the spectrum through yellow to yellowish-green, we can see many shades of colour, but a bee sees only yellow. Similarly, a bee can see one colour in the greenish-blue wave-band, and one in the blue wave-band. Out of all the different shades of colour that make the world so beautiful to us, the bee can see only three.

Yet a bee is ahead of us in one respect. I said that a bee can distinguish *four* colours. The fourth is in the ultra-violet part of the spectrum, which to us is invisible. What 'colour' the bee sees in the ultra-violet I am afraid we shall never know. Even if one could ask a bee—and I would not put it past the ingenuity of some future research worker to find some means of communicating with these remarkable insects—it could not tell us, any more than we could tell a person blind from birth what 'red' looks like.

If the bees are one up on us with their ultra-violet colour vision, we have the advantage in another respect. Bees cannot see red at all. A bee will confuse all colours nearer than orange to the red end of the spectrum with dark grey. This does something to restore the prestige of man.

If the conclusions reached by Von Frisch are correct, we should be able to apply his findings to the colours of the flowers that are commonly visited by bees. When we do so we find that it works. Most bee flowers are blue, yellow or purple. Even the exceptions confirm Von Frisch's discoveries. The poppy with its bright red flowers ought

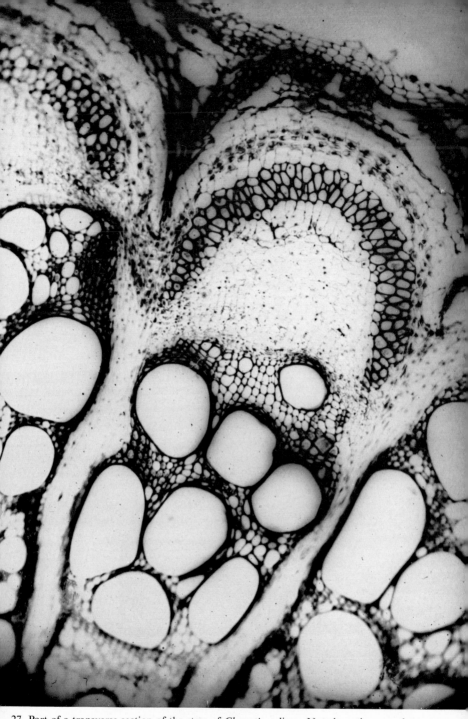

27. Part of a transverse section of the stem of *Clamatis*, a liane. Note how the secondary wood is divided into segments by thin-walled cells, for flexibility

28.
The stag-horn fern
(*Platycerium*), a tropical
fern that grows as an
epiphyte on the trunks of
trees. Royal Botanic
Gardens, Kew

29.
The sundew plant. Note
the tentacles growing
from the leaves. Wisley
Heath, Surrey

not to be noticed by bees, yet they visit poppies frequently. On further examination, we find that the red petals of the poppy reflect a great deal of ultra-violet light. We cannot see this, but the bees can.

Many other flowers reflect in the ultra-violet, and this may explain why some of them that are inconspicuous to us are notably attractive to bees. The flower of the white bryony (*Bryonia dioica*) is one. It has small flowers of a greenish-yellow colour that to us do not show up clearly against the foliage. To bees, however, they are conspicuous.

Besides the colour of a flower, we also have to consider its form. Von Frisch showed the bees were readily able to choose between different shapes, such as a star and an oval, but colour came first. If bees were trained to come to a blue star, and were then given the choice between a yellow star and a blue oval, they chose the blue rather than the yellow in spite of the change in shape.

Later, Hertz showed that bees would choose broken patterns rather than unbroken ones of the same area. This indicates that they will prefer loose groups of small flowers to a single large flower.

Colour and scent

The colour of flowers is due to one of two causes. Most yellow flowers and some red ones contain structures called chromoplasts that are coloured by carotenoid pigments. Other flowers, especially blue ones, contain pigments called anthocyanins dissolved in the cell sap. Anthocyanin pigments change colour according to the reaction of the cell sap, turning from red when the sap is acid to blue or violet when it is alkaline. This explains why flowers such as the comfrey (*Symphytum officinale*) vary from pink to purple, even on the same plant.

Colour in other parts of a plant may depend on the same factors. The red colour of ripe tomatoes and the pink roots of the carrot result from the presence of chromoplasts in the cells, while the dark red of the beetroot and the richly coloured leaves of the coleus are the result of anthocyanin pigments. The colours of autumn leaves are partly the effect of decomposition products of chlorophyll, and partly due to the development of anthocyanins in the ageing leaves.

Besides attracting bees by their colour, most bee flowers are scented. Von Frisch investigated the effect of different scents on bees, and was the first to prove that their scent receptors were situated on their antennae; bees from which the antennae were removed had no

129

sense of smell. He tried the effect of training bees to associate different scents with food, and showed that they were able to discriminate between a wide variety of smells. He also showed that bees could be trained to come to scents that were entirely artificial in origin, such as lysol and carbon disulphide, the latter having a smell reminiscent of rotten cabbages.

Von Frisch also experimented on the relative importance of scent and colour in attracting bees. He trained bees to come to a blue card scented with jasmine, after which he gave them a choice of a yellow card with jasmine and a blue card without scent. Most of the bees went to the yellow card, which smelt right but was the wrong colour. Then he placed a screen a yard long between the two colours, so that the bees had to make up their minds which card to fly to while they were still at a distance. In this experiment, most of the bees flew to the blue card. He concluded that colour was the stronger stimulus at a distance, though scent was more effective when the bees were near to the object.

This conclusion by Von Frisch has proved to be correct up to a point, but only to that limited extent. Owing to the structure of their eyes, bees are very short-sighted by our standards. They cannot see a flower from a distance greater than a few feet. When they come to flowers from afar, therefore, they must be guided by scent and not colour. This is certainly the case with a crop such as clover, where great masses of flowers are grouped together; there can be little doubt that their combined scent is sufficient to attract bees. With individual flowers, however, it is doubtful if the scent is strong enough to attract bees when the flower is out of their visual range.

There seem to be three stages in the attraction of bees to flowers. From a distance they may be attracted by scent, if it is strong enough, which gives them the general direction in which to fly. When they are within visual range of the flowers—that is, within a yard—scent is no longer important; the visual sense takes over, and they 'home' on the flower by sight. When the bee is close to the flower, however, scent again becomes the dominant sense, for a bee will not alight on a flower unless it smells right.

The sensitivity of a bee to smell is a matter of argument. Von Frisch concluded from his work that the sense of smell in a bee is of about the same sensitivity as in a man, but later work, using more refined methods, has shown that bees have much more sensitive 'noses' than we have. It is now thought that a bee has a sense of smell not less than ten, and possibly a hundred times as strong as ours.

Scent in flowers is usually produced by the presence of volatile essential oils, which are mixtures of hydrocarbons (substances composed of hydrogen and oxygen, and nothing else) called 'terpenes', and their oxygen derivatives. Familiar examples are lavender oil from the lavender (*Lavendula vera*), oil of aniseed from the aniseed (*Pimpinella anisum*) and peppermint oil from the peppermint (*Mentha piperita*). Many are of commercial value for making perfumes. Essential oils are also responsible for the flavours of spices such as cloves (the fruit of *Eugenia caryophyllata*), ginger (the rhizome of *Zingiber officinale*), pepper (the berries of *Piper nigrum*) and cinnamon (the bark of *Cinnamomum zeylanicum*). The hop (*Humulus lupulus*), which gives the bitter flavour to beer as well as acting as a preservative, contains essential oils called 'lupulins' that are produced by hairs on the bracts of the clusters of female flowers.

What I have just said about the senses of bees refers particularly to the hive bee, on which most of the work has been done, but there can be little doubt that it applies also to solitary bees, such as the bumble bee. The hive bees have one important guide which the others lack: they send out a few bees as 'scouts' to find suitable places for foraging, and when the scouts return to the hive they 'tell' the other bees, by means of a complicated dancing routine, in which direction to go. The bumble bee, being solitary, has to depend on its own resources, yet in spite of this it appears to do very well.

Honey guides and nectar

We now come to the part played by the markings on the flower known as honey guides—the 'Saftmale' of Sprengel. These are very common, especially in the more highly developed types of flower. They often consist of lines that lead towards the nectary, as in the sweet violet (*Viola odorata*), which has pronounced streaks on its lower petal, but they may take the form of spots on the corolla tube, as in the foxglove (*Digitalis purpurea*); or the centre of a flower may be of a different colour from the rest, as is the yellow centre of the forget-me-nots (*Myosotis*).

Sprengel considered that these markings guided the bee towards the concealed nectary of the flower, and subsequent work seems to show that he was right. When a bee alights on a flower it usually touches down on the edge of the corolla, apparently attracted in the first instance by the colour contrast between the flower and its background. Where honey guides are present, however, the insect is stimulated by the contrast between the colour of the guide and the rest of the flower. Instead of alighting at the edge of the flower, it comes down nearer to the centre, and so has a better chance of finding the nectary.

There are, of course, plenty of bee flowers without honey guides of any kind, and one is tempted to wonder how the bee manages in their absence. But bees have a remarkable capacity for learning from experience, and after a few trials on a flower without honey guides, they soon learn where the nectaries are. Moreover, there may be honey guides that we cannot see. Apart from the ability of the bees to see ultra-violet light, it has been shown that in some flowers the scent is not evenly distributed all over the petals. Some flowers have their honey guides marked out in scent instead of colour. This may also happen where visible honey guides are present, for it has been shown that the guide lines may have a different smell from the rest of the flower.

Structure of pollen flowers

Bees visit flowers in order to take pollen as well as nectar, and there are plenty of bee-pollinated flowers that are devoid of nectar; these are the 'pollen flowers'. One is the common broom (*Sarothamnus scoparius*). This flower also shows one of the many mechanisms for securing cross-pollination. It has the characteristic butterfly shape of the Papilionaceae (pea and bean family). The upper petal is large and stands upright; it is known as the standard, and bears honey guides in the form of dark lines, although it is a pollen flower. The two petals at the side of the flower are called the wings, and the two lower ones are united to form the keel, a boat-shaped structure which encloses the androecium and gynoecium. The edges of the keel petals interlock with those of the wings. The androecium consists of ten stamens, joined together by their filaments. Five of the stamens are long and five are short. The gynoecium consists of a single carpel, with a long ovary and a long style, with the stigma at its tip. The stamens and style are packed inside the keel, and are in a state of considerable tension.

When a bee—usually of a heavy type, such as a bumble bee—alights on the wings of the flower its weight upsets the interlocking of the wings with the keel, allowing the edges of the keel to separate,

132

the separation beginning at the inner part of the keel and progressing towards the tip. As the two petals of the keel come apart the five shorter stamens are the first to be released from confinement; they spring out and strike the bee on the lower side of its body, throwing out their pollen. As the opening of the keel continues, the longer stamens and the style are finally released. They spring out violently, pollen from the stamens being thrown on to the back of the bee. The style is under even greater tension than the stamens, so that it springs out a little faster and touches the bee slightly before the stamens shed their pollen. If the bee has previously visited a flower of the broom, the stigma will pick up pollen from it. After touching the bee the style continues to curl, so that the stigma is carried out of the way of the stamens when they discharge their pollen.

The explosive mechanism of the broom flower is irreversible, for once the keel has opened to release the stamens and style they cannot be folded back inside it, nor can the edges of the wings and keel again be made to interlock. On a broom plant you can recognize the flowers that have been pollinated because their petals hang limply, and such flowers will afterwards be found to set seed, while flowers that have not been visited by bees do not. The mechanism seems to be successful in avoiding self-pollination.

The pollination mechanism of the closely related gorses (*Ulex*) is similar to that of the broom, but is less explosive. When the weight of the bee falls on the wings the keel opens and the style emerges first and touches the lower side of the body of the bee; the stamens appear afterwards, dusting the lower side of the insect's body with pollen. Since the style is first to leave the protection of the keel, cross-pollination is even more certain in the gorse than in the broom.

Flowers that provide nectar

A quite different kind of bee pollination mechanism is seen in species of *Salvia*, plants belonging to the mint family (Labiatae). The five petals are united to form a tube, though their upper ends are free and form a series of projecting lobes or lips. The upper lip, which consists of two fused petals, stands erect, its concave interior housing the two stamens and the young style. The lower lip of the corolla forms a landing-stage for bees. The ovary is hidden at the bottom of the corolla tube, and has a nectary at its base. The stamens are constructed in a most peculiar way. Their anthers are in two separate halves, the lower half of each being sterile and containing no pollen. The two halves of the anthers are joined, in each stamen, by a long bar called the 'connective', and the connectives are curved so as to fit inside the concave upper lip of the corolla. The connectives are joined to the filaments of the stamens near their lower ends, and the joint is not a rigid one; if the lower halves of the anthers, which lie in the mouth of the corolla tube, are pushed hard enough the connectives can swing, causing the upper fertile halves of the anthers to drop down.

When a bee alights on the landing stage formed by the lower lip of the corolla it pushes its head into the corolla tube to get at the honey

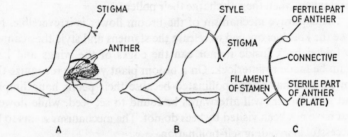

Fig. 26. Flower of *Salvia*, showing the pollination mechanism.
A, a bee, on visiting a young flower, pushes against the lower, sterile part of the anther, swivelling it on its joint, so that the upper, fertile part is brought forward and downward, touching the bee's back and dusting it with pollen. B, an older flower, in which the stamens have withered, and the style has grown, bringing the stigma into a position where it will touch the back of a visiting bee and gather pollen. C, a stamen, showing how the long connective is mounted on the short filament

secreted by the nectary at its base. The lower, sterile, halves of the anthers are in its way, but by pushing against them the bee makes them swivel, thus bringing the upper fertile halves of the anthers down to touch its back, which is dusted with pollen. (Fig. 26).

At this time the stigma is still young, and is not receptive to pollen. In older flowers which by this time have shed their pollen, the style grows longer and bends downwards. This brings the stigma, which is now receptive, into a position where it will touch the back of a bee visiting the flower. If the bee has come from a younger flower it will be carrying pollen on its back, and some of this will be transferred to the stigma.

Besides the remarkable swinging stamens, *Salvia* is provided with another mechanism for avoiding self-pollination. The pollen and the stigma mature at different times. This is called 'dichogamy', and it is very common in plants with hermaphrodite flowers (flowers with both stamens and gynoecium). In *Salvia* it is the stamens that mature first, a condition known as 'protandry', which is very widespread; we find it in the rosebay willowherb (*Chamaenerion angustifolium*), nearly all the Compositae (daisy family) and Umbelliferae (cow parsley family), and many others too. The opposite to protandry is 'protogyny', where the stigma ripens before the pollen. This is less common, but it is found in many plants, including the plantains (*Plantago*), the meadow rue (*Thalictrum flavum*) and the bear's foot (*Helleborus foetidus*).

Dichogamy is extremely common in wind-pollinated flowers. One might expect this. In such a casual process as wind pollination it is hard to think of any other way, other than by the sexes being separate, in which self-pollination could be avoided.

Another device for the avoidance of self-pollination is 'heterostyly'. This means that there are two or more different kinds of flower, with stamens and styles of different lengths. The commonest example is the primrose (*Primula vulgaris*). If you examine a collection of primrose flowers from different plants you will find that they are of two kinds; pin-eyed and thrum-eyed. A pin-eyed flower has the stigma in the mouth of the corolla tube; you can see it as a little round spot when you look straight into the face of the flower. You cannot see the stamens, for the anthers are situated half-way down the corolla tube and so are hidden by the stigma. A thrum-eyed flower, on the other hand, has the anthers of the stamens visible in the throat of the corolla, the stigma being set about half-way down the tube, in the position occupied by the stamens in a pin-eyed flower (Fig. 27).

The primrose is pollinated by the honey bee. When the bee inserts its long proboscis into the corolla tube of a thrum-eyed flower to get at the nectar it will receive pollen at the right level to pollinate the stigma of a pin-eyed flower, and vice versa. As an additional precaution against self-pollination, the pollen grains of a thrum-eyed flower are larger than those of a pin-eyed flower, while the stigma of a pin-eyed flower has coarser papillae than those of a thrum-eyed flower, so that the larger thrum-eyed pollen sticks better than the small pollen grains from a pin-eyed flower.

The flower type in the primrose has been shown to be controlled by a single gene, thrum-eye being dominant to pin-eye. About seventy per cent of primroses are thrum-eyed, the rest pin-eyed.

In the purple loosestrife (*Lythrum salicaria*) heterostyly is carried a stage further, for there are three kinds of flower, which have long, short and intermediate style and stamen lengths. A flower with a long style has both short and intermediate stamens, and so on. Any one type can pollinate either of the other two.

Some of the most remarkable mechanisms for insect pollination are found in the orchids. The flower of an orchid shows extreme specialization, and in many respects it is quite unlike the flower of any other plant. There is a single stamen and the pollen grains are stuck together, forming a structure called a 'pollinium' in each

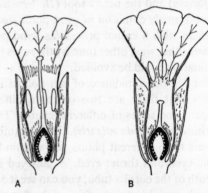

Fig. 27. Heterostyly in the primrose.
A, a pin-eyed flower and B, a thrum-eyed flower

anther. The pollinium is transferred as a whole by the bee. When the pollinia are mature, they are usually attached to a little pouch, the 'bursicle', just above the stigmatic surface, ready to adhere to a visiting bee (Fig. 31).

I have described only a few out of the multitude of ways in which flowers can be adapted for bee pollination. The bee is not the only insect that pollinates flowers, and some of the ways in which flowers are adapted for pollination by other insects are every bit as remarkable as the adaptations for bees.

Pollination by wasp, moth, beetle and fly

One of the most extraordinary ways in which a plant can become adapted to an insect is shown by the fig, which is pollinated by a wasp, *Blastophaga grossorum*. The wild fig (*Ficus carica*) occurs in Italy. Its 'flower' is really a hollow inflorescence called a 'synconium', pear-shaped, with an opening at its apex, and the flowers are grouped inside it. Just inside the mouth of the synconium are the male flowers (the fig is a plant which has separate sexes), while farther in are the female flowers. The fruit of the fig is a false fruit, consisting of the swollen synconium with the true fruits (the pips) inside it (Fig. 28).

The wild fig, or 'fico silvatico', bears its first crop of flowers in early spring. Numerous male flowers are formed at the mouth of the synconium, and farther in are what are called 'gall flowers'. These are female flowers with single seeds, which cannot form edible fruit.

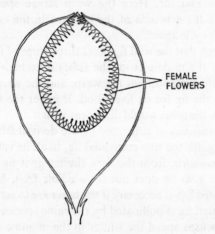

FEMALE FLOWERS

Fig. 28. A fig, in vertical section

Female wasps enter the synconia and lay their eggs in the ovules of the gall flowers, one egg being deposited in each. Male wasps which hatch from the eggs spend their whole lives inside the synconium; they complete their metamorphosis in the ovules and, emerging as adult wasps, seek out gall flowers containing female wasps. Piercing the wall of the ovary they fertilize the female wasps; after which, their work completed, they die without leaving the synconium in which they were born.

The fertilized female wasps now leave the synconium, and as they do so they have to crawl over the male flowers near its mouth, becoming dusted with pollen as they do so. By this time the first fruit of the fig, which is known as the 'profico', is ripe, though it is too bitter to be eaten. The female wasps fly very little; they crawl about the branches of the tree, looking for the young inflorescences that compose the second crop of wild figs, in late spring. These are synconia, resembling those of the first crop except that they contain only female flowers, this time with fully developed flowers that are capable of setting seed. The wasps try to lay eggs in the ovaries of these flowers, but are frustrated, and while they do so they pollinate the flowers. The fruits or 'fichi' ripen in September, and are edible.

While the fichi are ripening, a third crop of fruits is beginning to form. These are the 'mamme', and they contain gall flowers only. The wasps, having failed to lay their eggs in the fichi, do so in the gall flowers of the mamme. Here the wasp larvae spend the winter, emerging from the synconia of the mamme in the early spring and starting the life cycle again.

It will be seen that the wild fig bears three crops of fruit every year, of which only the middle crop (the fiche) is edible. The fig cannot complete its life cycle without the wasp, and the wasp is completely dependent on the fig for its livelihood. If either the fig or the wasp ceased to exist, the other would disappear with it.

The many varieties of cultivated fig are derived from the wild fig. We can distinguish the true cultivated fig, in all its varieties, as *Ficus carica* var. *domestica*, from the very similar goat fig (*F. carica* var. *caprifica*). The goat fig does not form edible fruit, but its presence among cultivated figs is necessary if the latter are to set fruit.

The cultivated fig is pollinated by the same species of wasp as the wild fig. The wasps spend the winter in the mamme of the goat fig, and in early spring they emerge from the mamme and enter the profichi of the goat fig, as well as the 'fiori di fico' of the domestic fig. The profichi of the goat fig contain gall flowers in which the insects lay their eggs, but the fiori di fico contain only sterile female flowers in which egg-laying is impossible. The fiori di fichi usually fall off the tree without ripening into fruit, but in some varieties a crop of fruit may be set.

The female wasps emerge from the profichi of the goat fig in early summer, becoming dusted with pollen as they escape. They then have a choice of two kinds of synconia: the 'pedagnuoli' of the domestic fig and the 'mammoni' of the goat fig. Those that enter the pedagnuoli find functional female flowers in which they cannot lay their eggs, although they pollinate the flowers with pollen from the goat fig. Those that enter the mammoni, on the other hand, find gall flowers waiting for their eggs.

138

The pedagnuoli ripen and form the main crop of edible figs. In the mammoni of the goat fig, which are not edible, the eggs hatch and the new generation of female wasps is fertilized by the males, just as in the profichi of the wild fig. The female wasps that escape from the mammoni are dusted with pollen from male flowers near the exit, and they in turn have a choice of two destinations. Some enter the 'cimaruoli' of the domestic fig, where they find functional female flowers in which they cannot lay their eggs, though they pollinate them with pollen from the mammoni of the goat fig. The cimaruoli may develop into a winter crop of figs. Others enter the mamme of the goat fig, where they lay their eggs in the gall flowers. Here the eggs hatch, and the larvae pass the winter, the wasps emerging in the spring, when the life cycle starts afresh.

The domestic fig bears only female flowers, and so is absolutely dependent on pollen from the goat fig for the setting of fruit. It is also absolutely dependent on the wasp for pollination. This fact is recognized by the peasants of Southern Italy, who always grow goat figs among their domestic figs. When the fig was introduced to California it would not bear fruit until the goat fig and the wasp *Blastophaga* were introduced with it. In the north of Italy the domestic fig is grown alone, varieties being chosen that set fruit without fertilization ('parthenocarpic' varieties), so that 'caprification' is unnecessary.

From a botanical point of view the story of the fig and its pollination is of particular interest because we have an instance of a monoecious plant (unisexual flowers on the same plant) which has lost its male flowers on domestication. This is unique in the plant kingdom. If seeds of *Ficus carica* var. *domestica* are sown they revert to the wild type, so that the domestic fig can only be propagated by cuttings or graftings.

Many flowers are pollinated by moths, familiar examples being the honeysuckle (*Lonicera periclymenum*), the evening primroses (*Oenothera*) and the sweet-scented tobacco (*Nicotiana affinis*). Moth-pollinated flowers have certain characteristics in common: they have long corolla tubes, because moths have very long 'tongues'; they tend to be white or light-coloured, showing up well in the dark; and they usually emit a strong scent in the evening, when moths are about. Usually there is no landing stage provided in lepidopterous flowers as there is in bee flowers, for butterflies and moths usually flutter while sucking nectar, instead of coming to rest on the flower.

One of the most remarkable instances of the mutual dependence of an insect and a plant is seen in *Yucca filamentosa*, a South American plant belonging to the lily family (Liliaceae). This is pollinated by a particular species of moth known as *Pronuba yuccasella*. This moth, like all members of its genus, is provided with peculiarly spinous, prehensile maxillary appendages. The *Yucca* flower emits its scent at night, when the moth visits it, going first to the stamens where it collects a quantity of pollen, uses its maxillary appendages to knead it into a firm ball about three times as big as its head, and carries it away. It then flies to the ovary of another flower, where it lays its eggs among the ovules; it has a long ovipositor which enables it to do this. Then it climbs down the style of the flower, which hangs downwards, and on reaching the stigma it presses the ball of pollen into the grooves on the stigma, ensuring that the flower is properly pollinated.

After the flower is fertilized, the fruit develops. The caterpillars of *Pronuba* feed on the ovules as they develop into seeds. The ovules of *Yucca* are so numerous that there are plenty left to form seed even after the caterpillars have eaten their fill. The *Yucca* plant and the *Pronuba* moth are so adapted to one another that neither can live alone; *Yucca* when grown in Europe, where the moth does not occur, never sets fruit.

The hover flies (Syrphidae) are also good pollinators, and some flowers are adapted to them. A common one is the germander speedwell (*Veronica chamaedrys*), very common by roadsides and in waste places. It has small blue flowers with four petals (the upper one is really two fused together), two stamens which hang downwards, and a fairly long style which hangs down between the stamens. The petals at first sight appear to be free from one another, but closer inspection reveals that they are united at their bases to form a short corolla tube, at the bottom of which is a nectary.

When a hover fly settles on a flower of *Veronica* the stigma touches its underside and receives pollen from another flower. In getting at the nectar, the insect grasps the two stamens with its forelegs, drawing them together and becoming dusted with pollen as it does so. The hover fly carries the pollen to another flower, where the sequence of events is repeated. The flowers are markedly protogynous, which reduces the risk of self-pollination.

The hover fly belongs to the class Diptera, or true flies, to which the house fly also belongs. Many other diptera pollinate flowers, and certain floral characters are often associated with fly pollination, such as a dirty brownish, greenish or purplish colour, and a foetid odour.

The wild arum, cuckoo-pint, or lords-and-ladies (*Arum maculatum*) shows a high degree of specialization for pollination by flies. A large sheathing bract is curled round the central axis that bears the flowers, and is constricted at its lower end, forming a 'waist'. The top of the floral axis or 'spadix' is prolonged into a club-shaped spike which is of a reddish colour. Lower down are the flowers, which are in three groups. Lowest of all are the female flowers; above them come the male flowers, and at the top is a group of abortive male flowers which are modified to form a ring of downward-pointing hairs. These hairs are placed just where the bract narrows to form its waist, and so partially block the passage that leads down to the functional flowers (Fig. 29).

Small flies, especially midges, are attracted to the inflorescences by the evil smell, enticed too by the red spadix. They crawl down inside the bract, being able to push their way past the zone of hairs. They cannot escape, however, because the downward-pointing hairs bar their passage, and because a slippery oil secreted by the cells of the lower part of the axis adds to the difficulty of climbing up.

At this time the stigmas of the female flowers are receptive to pollen, and if the insects are carrying pollen from a previous captivity the flowers are pollinated. Later the male flowers mature, the hairs that block the way out shrivel, the slippery oil dries up, and the insects are able to escape. As they do so they are sprinkled with pollen as they crawl over the male flowers. Not satisfied with having been caught once, the flies are liable to visit another inflorescence and pollinate a further set of female flowers with the pollen that they are carrying on their bodies.

The wild arum cannot pollinate itself because the stigmas of the female flowers have withered before the stamens of the male flowers have produced their pollen.

In temperate regions flowers pollinated by flies are usually small and insignificant, though they may have a strong and unpleasant scent. In the tropics it is different. Here the fly flowers are among the biggest. *Rafflesia arnoldi*, a parasitic plant from Sumatra, has flowers a yard across. They lie on the ground springing direct from the roots of the host, which is usually a tree. There are many tropical aroids which attract flies and beetles in much the same way as the cuckoo pint. One aquatic genus, *Cryptocoryne*, has the tubular bottom of the bract drawn out into a long tube down which beetles slip into a chamber eighteen inches below the surface of the water.

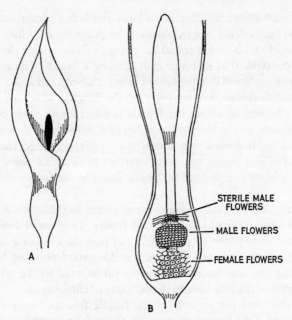

STERILE MALE
FLOWERS

MALE FLOWERS

FEMALE FLOWERS

A

B

Fig. 29. The wild arum. A, view of the spathe, with the red tip of
the spadix showing. B, enlarged view of the spathe, with
the bract cut away to show the spadix

Beetle flowers are not uncommon, especially in the tropics. They include the flowers of *Magnolia* and the water lilies. The gigantic *Victoria amazonica*, with floating leaves six feet across and flowers up to fifteen inches wide, is pollinated by cockchafers, which eat the stamens and are then trapped in the gigantic flowers, which close up in the morning; in the evening, when the flowers open again, the beetles are released. Perhaps 'trapped' is not a good word here: the cockchafers, which are nocturnal insects, may appreciate the warmth and comfort of the closed flowers to lie up in during the daytime, not to mention the protection from their enemies that the flowers afford them.

Beetle-pollinated flowers are usually of the open, primitive type, with little specialization about them. This has led to the very reasonable theory that beetles were the first insects to pollinate flowers, pollination by more specialized insects such as bees coming later. The beetles are an old group of insects—much older than the bees. They are clumsy pollinators, usually feeding on the stamens and

142

breaking those they do not eat. When the bees appeared on the scene, with their long tongues for sipping nectar delicately, a far more subtle technique became possible.

Pollination by birds and bats

In warm countries such as Brazil, South Africa, South Australia and Patagonia many flowers are bird pollinated, the birds being mainly humming birds (Trochilidae) and honey-creepers (Coerebidae) in America; honey eaters (Meliphagidae), and brush-tongued parrots (Trichoglossidae) in Australia; and a variety of different birds, including the nectar birds (Nectarinidae) and flower peckers (Dicaedae) in Africa. The characteristics of a bird flower are brilliant colour, absence of scent, a strong, elastic structure and copious supplies of rather watery nectar. This last point is particularly noticeable in many bird flowers; the American *Erythrina cristagalli* is known as the cry-baby flower because of the nectar that drips out of its inverted inflorescence. In Java, the flowers of *Erythrina* are the main source of fluid during the dry season for birds as large as pigeons. The nectar of some Australian species of *Protea* is produced in such quantities that the aborigines collect it.

Many bird flowers are bright scarlet. This is not surprising when we consider the eye of a bird. Most birds have a yellowish oil in their retinas which acts as a kind of colour filter, giving the effect of seeing the world through pale orange spectacles. This makes the birds colourblind to blue, though they can distinguish green and (particularly) red. In 1915 a list of 159 'ornithophilous' (bird-pollinated) flowers was published, of which eighty-four per cent were red.

Birds usually sip nectar while fluttering, so bird flowers are usually not modified, as are bee flowers, to provide a landing stage. The South African *Strelitzia reginae* is an exception, with a perch that consists of a modified bract.

Birds are clumsy creatures where flowers are concerned—even more clumsy than the beetles, as they tear the blossoms and break the stamens. It is not surprising, therefore, that in many ornithophilous flowers we find the delicate ovary concealed in the floral receptacle ('inferior ovary'), and stamens with woody filaments that resist breaking.

Which came first: bird pollination ('ornithophily') or insect pollination ('entomophily')? We do not know for certain, but there is a very plausible theory that bird pollination prevailed in Cretaceous times in the tropics, that later there was insect pollination in more temperate climates, and later still there was wind pollination ('anemophily').

Pollination by bats ('cheiropterophily') is a natural sequence to ornithophily, and many nocturnal flowers in the tropics are bat pollinated, including the calabash tree (*Crescentia cujete*), the sausage tree (*Kigelia*) and the midnight horror (*Oroxylon*). Certain cacti of the genus *Cereus* are pollinated by vampire bats.

Bats damage flowers severely with their claws, and so bat flowers are large and strong. Like the bird flowers, they are copiously supplied with nectar. They usually open in the evening, and their colour tends towards greenish-yellow, brownish or purple. Their smell is peculiar, and usually rather unpleasant.

Snail pollination ('malacophily') has been reported for a few flowers, including the golden saxifrage (*Chrysosplenium alternifolium*). The evidence for snail pollination is rather dubious; it is more likely that the snails merely feed on the flowers.

Wind pollination

Many flowers are pollinated by wind, which is also the sole agent for the pollination of the conifers and their allies. Wind-pollinated (anemophilous) flowers usually have quite different characteristics from those pollinated by insects or birds. Large showy petals are absent, for they would only get in the way of wind-blown pollen without serving any purpose. Nectaries are absent, as also is scent. The flower has a very functional aspect, with no unnecessary fripperies (Fig. 30).

On the positive side, wind-pollinated flowers usually produce an abundance of pollen. Anemophily is a chancy affair, and for every pollen grain that finds its target on the stigma tens or hundreds of thousands—millions in some instances—will be wasted. Grasses are wind pollinated, and anyone who suffers from hay fever knows how much grass pollen is carried in the air. In springtime, when the male cones of the pine and the fir are shedding their pollen, the surfaces of mountain lakes in Switzerland are covered with a bloom of yellow.

The pollen of anemophilous plants must float easily in the air, and to that end it is small, and the grains are dry, so that they separate easily. Entomophilous flowers, on the other hand, tend to

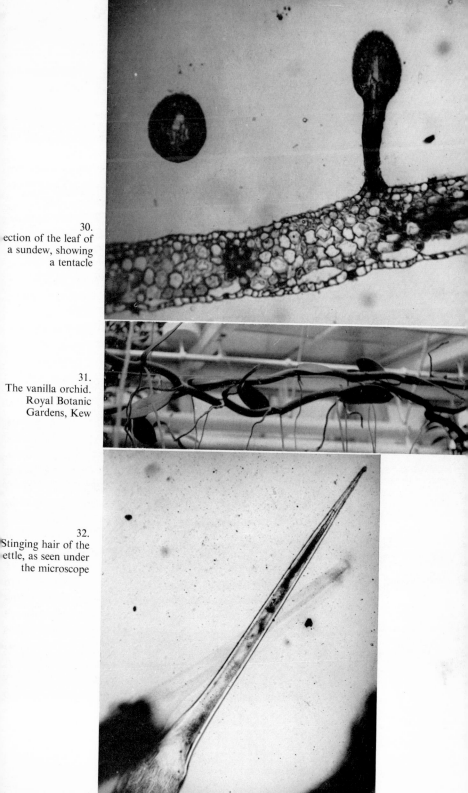

30.
ection of the leaf of
a sundew, showing
a tentacle

31.
The vanilla orchid.
Royal Botanic
Gardens, Kew

32.
Stinging hair of the
ettle, as seen under
the microscope

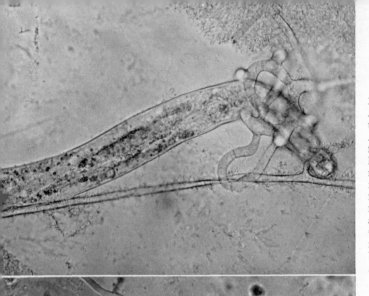

33.
An eelworm captured
the predacious fungus
Arthrobotrys robusta.
fore-end of the eelwo
has stuck to adhesive
networks shown on th
right of the picture. T
feeding hyphae of the
fungus can be seen, h
and there, in the bod
the eelworm. Greatly
magnified. (Photogra
by courtesy of *New
Scientist*)

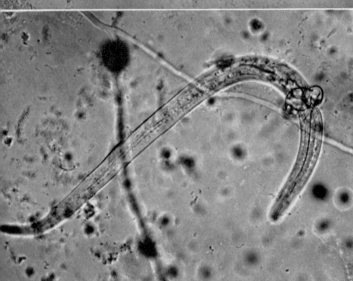

34.
Photomicrograph of a
eelworm captured by
predacious fungus
Dactylaria gracilis. T
eelworm has been ca
by passing its body in
one of the constrictin
rings carried by the
fungus; the swollen c
of the ring can be see
encircling the eelwor
(Photograph by court
of *New Scientist*)

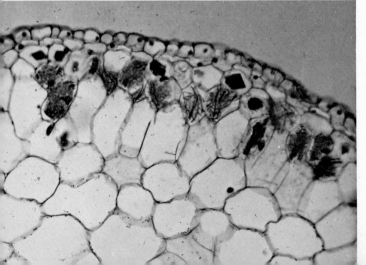

35.
Section of the root of
orchid, showing the
mycorrhizal layer in t
cortex

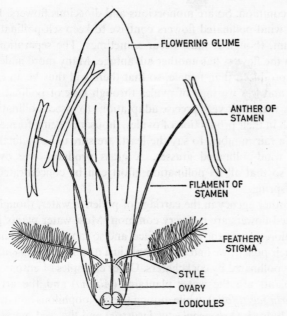

Fig. 30. A grass flower, removed from the inflorescence.
Note the long, feathery stigmas,
typical of a wind-pollinated flower

have large pollen grains that often stick together easily, forming yet bigger clumps. They have already booked their flight on a bee to the female plant, and it is an advantage, if as many as possible can travel together.

In an anemophilous flower the stamens are fully exposed, hanging out of the flower where the wind can readily get at them, and the anthers are often attached to the filaments by their centres by a swivelling joint so that the pollen is shaken out by any breath of wind. The stigmas, too, are conspicuous for their large size and feathery appearance; like the stamens, they are freely exposed when ripe. Notice how many of our wind-pollinated trees produce their flowers early in the year, before the leaves expand from the buds and get in the way of the pollen.

Wind being such an uncertain agent, scattering pollen broadcast without regard to where it falls, floral mechanisms to prevent self-pollination are lacking in anophilous flowers. On the other hand, dichogamy—the maturation of pollen and stigmas at different times—

145

is very common. So are monoecious and dioecious flowers. By these means, wind-pollinated flowers contrive to keep self-pollination to a minimum, though it must occur sometimes. The separation of the sexes in the flowers has another advantage. Many more male flowers can be produced than female, so that there will thus be an excess of pollen, and less wastage of ovules through lack of pollination. The orchids, with their very precise adaptations to insect pollination, can be lavish in their production of ovules, thousands being formed in one flower; a fair number, to say the least, are sure to be pollinated. Not so the wind-pollinated grasses. A grass produces one ovule per flower, so that all its pollination effort can be concentrated on an only offspring.

The other agency in the carriage of pollen is water, though water-pollinated flowers are not very common. Most water plants produce their inflorescences above the surface, and so can use wind or insects to carry their pollen; we had an example just now in *Victoria amazonica*, which is pollinated by cockchafers. Other examples of entomophilous water plants are the water plantains (*Alisma*) and the arrowhead (*Sagittaria sagittifolia*), while among the anemophilous aquatic plants are the bulrush (*Schoenoplectus lacustris*) and the reed maces (*Typha latifolia* and *T. angustifolia*).

Pollination by water

Some aquatic plants are water pollinated (hydrophilous) even where the inflorescences are at the surface. In *Vallisneria americana* the female flowers come to the surface at flowering time, carried up by the elongation of their stalks. The male flowers, which are minute, consist of two stamens within a perianth of three segments. They are detached from the plant under water and rise to the surface, where the perianth segments become reflexed back and support the flowers on the surface. They float about at random, and if they should approach a female flower the surface tension of the water draws them towards it. On touching, the male flowers tip over, so that the stigma is touched by their stamens. Should ripples in the water submerge the female flower the male flowers are enclosed with it in a bubble of air. When pollination has been effected the female flowers once more submerge, their stalks coiling up and drawing them beneath the surface, where the fruits are formed.

In the eel-grass (*Zostera marina*) we have a flowering plant that

has returned to the sea of its ancestors. This plant lives totally submerged in the sea, from low tide mark down to a depth of four yards
or so. The pollen is liberated under water and drifts in the water
currents to the stigmas of the female flowers. The pollen is filamentous,
and coils itself round the stigmas; it is thought that the shape of the
pollen helps it to float in water (see also page 110).

Self-pollination

Although most plants try to avoid self-pollination as a cat avoids sea
bathing, there are times when, for one reason or another, it becomes
desirable. Apart from plants such as the domestic wheat, which are
habitually self-pollinated, it may be that cross-pollination has failed
to take place. Rather than not set any seed at all, it is an advantage to
the plant to be able, as a last resort, to pollinate itself. Provision is
often made for this.

In the town-hall-clock or moschatel (*Adoxa moschatellina*) the
stamens stand at a lower level in the flower than the stigma, making
self-pollination unlikely. Later on, however, they grow up and touch
it. In the event of cross-pollination having failed, the flower will in
this way be self-pollinated. Many of the saxifrages behave in the
same way. In the yellow pimpernel (*Lysimachia nemorum*) and the
common St. John's wort (*Hypericum perforatum*) the stamens lean
well away from the stigma at first, but later they bend inwards and
touch it. In all these cases a chance is given for the flower to be cross-
pollinated before it resorts to self-pollination.

A more specialized way of securing self-pollination if cross-
pollination fails is seen in the 'cleistogamic' flower, which is not uncommon; the most familiar instance is that of the sweet violet
(*Viola odorata*). This plant flowers in spring, but the flowers, though
sweet smelling, are not conspicuous. Although they are visited by
insects, they seldom set seed. Later in the year, however, the plant
produces its cleistogamic flowers. These are hidden in the foliage and
never open, but they pollinate themselves, and usually set plenty of
seed.

Although it may be argued that cleistogamic flowers are produced
to ensure pollination when cross-pollination has failed to occur, this
point of view has not been proven, and there are valid objections to
it. It has been suggested that they may be formed under bad conditions
of growth, such as scarcity of food. The touch-me-not (*Impatiens*

147

noli-tangere), for instance, will produce cleistogamic flowers if grown in poor soil. Some plants produce cleistogamic flowers *before* their normal ones, and the wild clary (*Salvia horminoides*) nearly always forms cleistogamic flowers, normal open flowers being comparatively rare.

8 · The orchid plant

There has always been something mysterious and even sinister about an orchid. The blooms, besides being exotic and expensive, are unlike any other flower. One can with very little effort imagine them as being designed especially for the adornment of film stars and princesses. It seems absurd that they should serve the same function as a buttercup. Yet their floral parts are basically the same as those of other flowers, though they have become altered very much in the course of evolution, particularly with the object of securing cross-pollination by insects. Because of their unique form, and because the orchid family, Orchidaceae, is the largest of all plant families, they call for a chapter on their own.

The name 'orchid' is an ancient one, going back to the days of the ancient Greeks. Theophrastus, a student of Plato, is believed to have been the first to describe an orchid; at any rate, he found a plant with paired swollen roots that reminded him of testicles, and named it *orchis*, from the Greek word for a testicle. We cannot be sure whether this was really an orchid, for his description is meagre, but it probably was. The name *Orchis* is used today for one of the main genera of the orchids.

The resemblance of the tuberous roots of the orchids to testicles was noted in the middle ages, for the herbalists of the period collected 'dog stone roots', which, from their descriptions, were almost certainly the roots of orchids. They were much in demand at the time for making love potions. Orchids are still used medicinally in some parts of the world. The Zulus use preparations of *Habenaria* as an emetic; in Malaya, skin infections are treated with an extract of *Dendrobium*; and in South America they use *Spiranthes* as a diuretic and *Epidendrum* as a purge for removing tapeworms.

The orchids are monocotyledons, and there are altogether between twenty thousand and thirty-five thousand different species—nobody

149

can be quite sure what the actual number is. These are at present divided between about six hundred genera. They are all herbs, though some tropical species are climbers. Their distribution is worldwide, from the Arctic to the Antarctic, and they grow in almost every possible kind of habitat, from tropical rain forests to the tops of mountains, from bogs to deserts.

Epiphyte and terrestrial

Many of the tropical orchids are epiphytes—that is to say, they grow on other plants; not as parasites, but merely perching wherever they can find a root-hold. This is particularly so in the tropical rain forests, where epiphytic orchids can be seen growing on the branches of trees wherever one looks. Epiphytic orchids will even perch on telephone wires, somehow retaining their precarious hold and producing their flowers.

When an orchid is growing as an epiphyte its roots are of course out of contact with the soil. Most epiphytic orchids have long greenish or white roots that enable them to cling to their supports, sometimes in most unlikely positions, as in *Cattleya citrina*, which loves to grow upside down, with its roots in the air and its stem hanging towards the ground. Epiphytic orchids often have a curious development of the cortex of their roots, known as the 'velamen'. This consists of a zone of empty cells in the outer part of the cortex. The function of the velamen is still a matter of dispute among botanists, but the most general opinion is that the empty cells serve as a kind of sponge to mop up water. This is reasonable, for with no contact between roots and soil the obtaining of water must be something of a problem for the orchid plant, even in a tropical rain forest.

Orchids of temperate and colder regions of the world are mainly terrestrial, growing in the soil like other plants. It is among the terrestrial orchids that the tuberous roots from which they were named are prevalent.

Some orchids are saprophytes, their leaves reduced to scales that contain no chlorophyll, so that the plant must depend on rotting organic matter in the soil for subsistence. Saprophytic orchids usually grow in the litter of woodlands, where there is plenty of organic matter undergoing decomposition. Three British saprophytes are the bird's nest orchid (*Neottia nidus-avis*), the common coral root (*Corallorhiza trifida*) and the spurred coral root (*Epipogium aphyllum*).

These saprophytic orchids have an abundant underground system in which mycorrhizal fungi (see page 241) are particularly well developed. Threads from the fungus ramify throughout the surrounding woodland litter, digesting the organic matter by means of enzymes (ferments) and passing the products back to the orchid. In the bird's nest orchid the abundant roots form a tangle that is supposed to resemble the nest of a bird—hence the name. In the coral roots there is actually no root at all, for its place is taken by an underground stem or rhizome. In the common coral root the rhizome is broken up into many branches, each of which ends in several short secondary branches like knobs; it is the coral-like appearance of the rhizome that gives this orchid its name. The spurred coral root has a swollen rhizome bearing many rounded lobes, again resembling coral.

The bird's nest orchid and the spurred coral root are entirely without chlorophyll; the flowering shoot, which is the only part of the plant to rise above the ground, bears only colourless scale leaves. These two must obtain all the food they require from the woodland litter in which they grow, since they cannot manufacture the slightest morsel by photosynthesis. They depend on their fungal partners for everything. The common coral root, on the other hand, has a certain amount of chlorophyll in its flowering shoot, and so can make some food for itself, though not enough to live on.

In some orchids the saprophytic life is carried to fantastic lengths. V. S. Summerhayes, in his book *Wild Orchids of Britain*, mentions one Australian species where the whole plant is buried underground, flowering shoot and all. This orchid owes its discovery to the ploughing-up of soil that had formerly been undisturbed. Whether this orchid is pollinated by soil insects, or whether the flowers are cleistogamic, we do not know.

Another strange orchid is the Japanese *Gastrodia elata*, which appears to be a secondary parasite of oaks and other trees. The mycorrhizal fungus associated with this orchid is the honey agaric (*Armillaria mellea*), which is parasitic on trees, including the oak. This fungus is well known for its habit of producing thick underground threads called 'rhizomorphs', sometimes many yards long, by means of which it can spread from tree to tree. These rhizomorphs are also associated with the tubers of *Gastrodia*. The honey agaric absorbs its nourishment from the tree that is parasitized, and some of the nutriment extracted from the tree is used by the orchid.

Types of orchid flowers

It is in the structure of their flowers that the orchids are unique. Quite unlike that of any other plant, the flower shows extreme specialization for pollination by insects. It is the most advanced flower found among the monocotyledons, if not in the whole range of the flowering plants. Some are minute, measuring scarcely more than a millimetre in diameter, while others are a foot or more across. An orchid flower has three sepals that are petaloid—that is, coloured like petals. Inside these are three petals. The sepals and petals may be similarly coloured, or they may be of quite different colours. One of the sepals may differ from the other two in size or colour or both; this is always the upper one. The lowermost of the three petals is also modified, sometimes very much so, and the fantastic shapes that these lower petals can assume, as well as the remarkable colour patterns that they show, are largely responsible for the exotic appearance of many orchid flowers. Usually the lower petal projects from the flower, forming what is known as the lip or labellum, which may be divided into a number of lobes. The lip is often prolonged behind the flower to form a tubular structure called the spur (Fig. 31).

It is in the structure of the androecium and gynoecium that the orchid flower shows its greatest specialization. They are combined in a compound structure, the column, and on top is the single stamen, with two anther lobes containing pollen. It has already been said (see page 145), that in insect-pollinated flowers the pollen is often somewhat sticky, the grains adhering together in clumps so that an insect brushing against a stamen will carry away plenty on its body. In the orchids this tendency is carried to its logical conclusion, for in each anther lobe the pollen is compacted into a single club-shaped mass called a 'pollinium' (sometimes there are more than one). The pollinium is usually provided with a short stalk, the 'caudicle', by which, when mature, it is attached to a little pouch on the 'rostellum', a projecting part of the column just below the stamen. A visiting insect must remove the whole pollinium, which may contain thousands of pollen grains, at one go. To enable the comparatively massive pollinium to stick firmly to the insect, the base of the caudicle in many orchids is expanded into a flat sticky disc known as the 'viscidium'. In some orchids there is no viscidium, but an adhesive fluid from the rostellum sticks the pollinium firmly to the insect.

Below the rostellum are the stigmatic surfaces, numbering two in

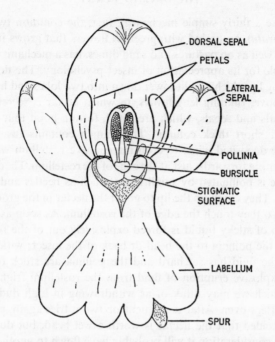

Fig. 31. Diagram of the flower of the early purple orchid
(*Orchis mascula*), face view

most orchids, one on each side of the column. They are often more or less joined together. The ovary is situated beneath the rest of the flower; it is very long, and often difficult to distinguish from the flower stalk. It contains many hundreds, often thousands, of ovules.

Another feature of orchid flowers that is not obvious to the casual observer is the phenomenon of 'resupination': a twisting of the whole flower during its development so that it finishes upside down. As the flower is first formed in the bud the labellum is on top of the flower, but as it develops the flower becomes twisted through an angle of 180°, so that when it finally opens the labellum is at the bottom.

Pollination mechanisms

The flowers of orchids are insect pollinated, and owing to their complex structure they are usually adapted to a particular kind of insect. The extreme adaptations shown by some of the orchids to ensure cross-pollination are quite fantastic, and no other flower can match them.

153

To take a fairly simple mechanism first, the common twayblade (*Listera ovata*), an orchid with greenish flowers that grows in beech-woods as well as on pastures and sand dunes, has a mechanism that is remarkable for its appreciation of insect psychology. The flower has a very long lip which is split lower down into two lobes and higher up has a groove, running lengthways, in which nectar is secreted. The other petals and sepals, which are small, form a hood that encloses the rather short thick column. The stamen contains two pollinia which are deposited when ripe on the concave rostellum with their bases pointing forwards, just at the edge of the rostellum. The common twayblade is pollinated by small insects such as beetles and ichneumon flies. They climb up the lip to get at the nectar in the groove, and in doing so they touch the edge of the rostellum. As soon as they do so a drop of sticky liquid is forced explosively out of the rostellum and fixes the pollinia to the head or back of the insect; within a few seconds the fluid has set hard and the pollinia are stuck fast. The sudden explosive eruption of fluid from the rostellum frightens the insect, which we may think of as withdrawing in high dudgeon to think matters over. After a moment or two it tries again, probably having decided that the nectar is worth a wet head, but during the pause for consideration it will probably have flown to another plant. If the flowers of the new plant are in the same stage of development as the other, it will again be dowsed and will collect some more pollinia. The twayblade, however, is protandrous—its stamen matures before the stigma becomes receptive. If the insect should visit an older flower it will find that the rostellum has bent upwards, exposing the stigma which will then receive the pollinia.

Observe that in this mechanism the sudden explosion of sticky fluid from the rostellum has two quite different effects. First it fixes the pollinia firmly to the insect, and at the same time it startles the insect into abandoning the flower and going to another one, which is usually on a different plant. In this way cross-pollination is made more likely.

A more specialized pollination mechanism is seen in the early purple orchid (*Orchis mascula*), one of our commonest British species, found in woods and open pastures mainly on lime-rich soils. Its spikes of purplish flowers are produced in May and June, and they are pollinated by bees. The upper sepal and the two lateral petals form a hood enclosing the column which is situated just in front of the entrance to the spur formed by the back part of the lip. At the tip of the column is the anther, containing two pollinia with bases drawn out into long stalks or caudicles, with sticky discs or viscidia. The pollinia are deposited by the ripe stamens on the rostellum, which has a minute purse-like structure called the 'bursicle' which receives the viscidia of the pollinia. The bursicle contains a sticky fluid which maintains the stickiness of the viscidia while the flower is awaiting pollination.

A bee lands on the lip of the flower and thrusts its tongue into the opening of the spur to get at the fluid contained in the cells that line its wall. In doing so its head presses down the flap-like side of the bursicle exposing the pollinia, one or both of which will stick to the upper part of its head. When the bee leaves the flower it bears the pollinia standing upright on its head like horns, firmly fixed by their sticky viscidia, the glutinous substance of which dries hard in a few seconds.

By the time the bee has flown to another plant, the position of the pollinia has changed. In a matter of about half a minute or so they bend forward through a right angle so as to lie parallel with the back of the bee, pointing forward like lances carried by a charging cavalry-man. When the bee visits the next flower the pollinia will miss the rostellum because of their position, passing beneath it and impinging on the stigmatic surfaces underneath. Some of the pollen will be broken off the pollinia and will stick to the stigmatic surfaces, so that the flower is pollinated.

Notice that the bending forward of the pollinia takes about half a minute. This gives the bee time to finish with the flowers on the spike where it collected the pollinia, and pass on to the next plant. Self-pollination is therefore most unlikely. If flowers are visited before the bending of the pollinia has taken place, the pollinia merely strike against the projecting edge of the rostellum and no pollination occurs. A bee may collect a dozen or more pollinia from one flower spike, and so will be able to pollinate many flowers before it needs recharging, so to speak.

It is easy to see this mechanism working if you can get hold of a spike of the early purple orchid or the spotted orchid (*Orchis fuchsii*), which is pollinated in the same way. Take a sharply pointed pencil and press it momentarily into the opening of the spur. You will see the pollinia standing upright on the end of the pencil, fastened by their viscidia. If you wait for a few seconds you will see them bend forward into their pollinating position.

To observe some of the most sophisticated of the orchid pollination mechanisms we must turn to the tropical orchids. In *Coryanthes* the lip is in the form of a bucket which is filled with a slightly intoxicating liquid. Insect visitors fall into the bucket and cannot climb out because the sides are slippery; things are not made any easier, perhaps, by the insect's being slightly bemused by the effects of the liquid in the bucket. There is one fairly easy exit, however, by a narrow opening just beneath the pollinia. This the insect takes, picking up the pollinia as it makes its drunken way out and transporting them to another plant.

A still stranger mechanism is seen in the South American orchid *Gongora maculata*. A bee landing on the lip in search of nectar finds it slippery and cannot keep its feet. It slides, willy-nilly, down the column which is curved and hangs downwards, and at the end of its ride it picks up the pollinia to be carried on its back to another plant. The same difficulties await it on the new flower, but this time the pollinia that the bee is carrying are brushed off on to the sticky stigma.

Pseudocopulation

The flowers of some orchids bear a striking resemblance to certain animals; in Britain, for instance, we have the bee orchid (*Ophrys apifera*), the fly orchid (*O. insectifera*), and the man orchid (*Aceras anthropophorum*). This mimicry is no doubt purely fortuitous in most cases, but sometimes it serves to promote pollination by attracting insects. In the mirror ophrys (*Ophrys speculum*), a Mediterranean species, and the small tongue orchid (*Cryptostylis leptochila*) of Australia, the form and colouring of the lip of the flower give it a close resemblance to the female of the pollinating insect. The flowers are visited by male insects which attempt to copulate with the 'females', and in doing so they either bear away the pollinia or deposit on the stigma pollinia that they may already be carrying. In each case the males of the insects concerned hatch earlier than the females, the orchid flowering during the period when the males are mateless. When the female insects appear the males are no longer interested in the orchids.

This phenomenon, called 'pseudocopulation', is probably quite common in those orchids that resemble insects. For the mechanism to work, the flowering of the orchid must be accurately timed to coincide with the period when males of the particular insect that is imitated are flying around in search of mates, and not finding them. Such accuracy seems to be well within the compass of the orchids.

It may happen that, in spite of the beautifully adjusted insect-pollination mechanisms of the orchids, cross-pollination does not succeed. Should this happen, many orchids have an emergency system by means of which they may be self-pollinated. This is all the more necessary in plants that are so precisely adapted for pollination by a particular species of insect, for in the event of insect numbers suddenly falling off through some chance twist of Nature, the normal pollination mechanism may fail. The means taken to ensure that self-pollination shall occur if cross-pollination fails vary in different species of orchids. In the bee orchid, for instance, as the flower grows old the column shrinks so that the pollinia are pulled out of the bursicle to hang down in front of the stigma, where any shaking of the flower will cause them to swing into contact with the stigmatic surface, thus securing self-pollination.

In the genus *Ophrys* the mechanism is a little more complex. When the flowers are young the column stands still and erect, but as the flower ages autolysis (self-digestion) begins in the tissues of the column, so that it sags downward. This brings the pollinia into contact with the stigma.

Germination of seeds

The changes that follow pollination are no less remarkable than the pollination mechanisms themselves. Orchid seeds are very small and are produced in enormous numbers. Those which have a million seeds in one fruit are quite common, and the capsule of *Cycnoches ventricosum* has been estimated to contain four million seeds. Darwin calculated that if all the seeds produced by a single plant of the spotted orchid were to grow into plants they would cover an acre of ground, and if all the seeds from these and succeeding plants throve and produced plants in their turn, in four generations the descendants of one spotted orchid would cover the whole of the earth. Perhaps it is just as well that mortality among orchid seeds is high.

In the orchids there is little or no development of endosperm in the seed, all the reserve food being in the embryo itself. Moreover, the embryo is not differentiated into radicle, plumule and cotyledon as in most monocotyledons but is just a small mass of undifferentiated cells. Differentiation of the embryo does not take place until after germination.

When an orchid seed germinates it first of all forms a swollen corm-like structure called a 'protocorm'. In this condition it stays, sometimes for as long as two years and normally will not develop any further until it has become infected with its appropriate mycorrhizal fungus, usually a species of *Rhizoctonia*. Then it continues to grow very slowly. From the top of the protocorm the first leaves make their appearance; later on, roots grow out from the bottom of the protocorm. The whole process is slow, and it will be many years before the plant is mature enough to bear flowers and set seed in its turn. For the mycorrhizas of orchids, see page 241.

The necessity for the mycorrhizal fungus to infect the protocorm before further development can take place accounts for the fact that orchid growers have found the seeds of orchids difficult to germinate successfully. It became customary to say that an orchid seed could not grow without its fungus. This extreme view was shown by Knudson to be false.

He took the view that the fungus might not play such a fundamental part and studied the possibility that the function of the fungus was merely to 'lend' the orchid plant certain enzymes that it lacked, thus enabling it to break down complex organic compounds to form simpler compounds such as sugar. He therefore put orchid seeds on to a special sterile culture medium containing sugar among other things, and succeeded in growing them without the fungus to maturity and flowering. The part played by the mycorrhizal fungus has not yet been fully elucidated. At first sight it looked as if Knudson had proved his point, and that all the fungus contributed to the partnership was a supply of useful enzymes. Further work has shown that this is unlikely. Some orchids cannot be grown on Knudson's medium though they will grow if supplied with sterile extracts of the fungus, or with vitamins. At present we do not know the full story.

9 · Plant dispersal

Plants are condemned to remain rooted in one spot all their lives; although the creeping palm may move by putting down from its crown new stems that take root in the ground, its advance is slow. Even plants such as the strawberry, with stolons which cover the soil in all directions and root at intervals, cannot travel far by this means alone. To secure dispersal to the ends of the earth, if need be, a plant must have some means of travelling more effective than anything that its vegetative body can devise. In flowering plants this end has been achieved by adapting either the fruit or the seed as the agent of dispersal.

A seed is primarily a resting structure, so designed that within its resistant coat the embryo plant can wait in safety until the winter—or, in some lands, the dry season—is past, and conditions are once more suitable for growth. A seed is a compact structure, ideally suited to travelling, for it is small and light and it needs neither food nor drink until it has reached its destination. A seed—or the fruit that contains it—can stow away on the fur of an animal, in a man's trousers turn-up, in the gut of a bird, or even in the hold of an aircraft, without the carrier being aware of what he is carrying. When it is finally dropped, perhaps a thousand miles from its place of origin, the seed is ready to grow into a plant, if climate and soil are suitable. If the weather is for the moment unkind, the seed can wait; there is no hurry.

Seeds are formed from the ovules of a flower, after fertilization. Within the tiny ovule the cells divide and form the embryo plant which consists of a stem rudiment with a pair of seed leaves (cotyledons) on either side—one cotyledon instead of two if the plant is a monocotyledon. Usually there is a root rudiment as well. As the seed develops, food material, consisting mainly of oil or starch with a little protein, accumulates in the cotyledons or in the extra-embryonic

159

tissue called the endosperm; this will serve to nourish the seedling plant on germination until it has formed chlorophyll in its leaves and can feed itself. Finally, the integuments of the ovule thicken and harden to form the seed coat within which the seed is protected until the time for germination has come.

Fruit formation

While the seed is developing, the ovary of the flower is forming the fruit. It swells, sometimes enormously, and at the same time undergoes such changes as may be necessary for its dual purpose of protecting the seed and assisting in its dispersal. The wall of the ovary may become soft and fleshy, as in the tomato, making the fruit attractive to birds. It may become tough, with a covering of hooked bristles which enable it to stick to the hair of animals that will transport it. This happens in the fruit of the goosegrass or cleavers (*Galium aparine*). The fruit may be small and hard, with a parachute of hairs to enable it to float in the air and be dispersed by the wind, as in the dandelion (*Taraxacum officinale*). As the fruit dries up on the parent plant, there may come a time when tissue tensions inside it may cause it to explode, scattering its seeds like shrapnel to fall where they will; this happens in the gorses (*Ulex*). There is considerable variety in the means whereby fruits achieve dispersal of their seeds.

Botanists distinguish between true fruits which consist of nothing but the modified ovary and its contents, and false fruits (pseudocarps) which include other structures besides the ovary. The tomato and the plum are true fruits, the fleshy part consisting of the swollen wall of the ovary. The strawberry, on the other hand, is a false fruit, for here the fleshy part that is so delicious is really the swollen receptacle of the flower, and not the ovary at all. The true fruits of the strawberry are the pips on the outside; each one is a complete fruit, consisting of a hardened ovary containing one seed. The apple is another false fruit: the fleshy part is again the swollen receptacle of the flower and the true fruit is the core. Blackberries, raspberries, and loganberries are really collections of fruits, for each little ball of which the 'fruit' is composed is a fruit in its own right. It has the same structure as a plum, and is called a 'drupe'. The blackberry is therefore a compound fruit, consisting of a collection of drupes.

What is commonly called a fruit, therefore, may be one of several different kinds of object, botanically speaking. Fortunately, one does not need to know the botanical nature of a fruit in order to appreciate it. When we enjoy the luxury of a ripe peach or a slice of iced melon we are only doing what Nature meant us to do, for fleshy, succulent fruits are intended to be eaten by animals—chiefly birds in Britain— so that their seeds, after passing unharmed through the gut, can fall to the ground with the dung.

There are two principal kinds of fleshy fruit (excluding, for the moment, false fruits such as the strawberry). These are the berry and the drupe. A berry has a wall (pericarp) which is fleshy throughout, as in the gooseberry and the tomato (it may seem strange to call the tomato a berry, but to the botanist it is a reasonably typical example). We can distinguish between the firm outer skin or epicarp, and the fleshy central portion or mesocarp which makes up the bulk of the fruit. In the centre of the fruit are the seeds, which of course are the pips. The pericarp of a drupe is in three layers, not two: the epicarp or skin, the fleshy mesocarp, and an extra inner portion, called the 'endocarp', which is hard and surrounds the seed (drupes are usually one-seeded, whereas berries may contain one or several seeds). The plum and the cherry are examples of drupes. A plum-stone is not the seed: it is the endocarp containing the seed. If it is carefully broken open, the seed will be found inside.

Many of the fruits of commerce are either berries or drupes, though they often have abnormal structures which give them their particular virtues. Oranges, lemons, grapefruit, limes and tangerines are all different species of the genus *Citrus* (family Rutaceae); their fruits are berries in which the epicarp and mesocarp are both leathery. The epicarp forms the skin, and it includes numerous oil glands in the form of cavities which become filled with an essential oil that gives the characteristic smell and flavour to the fruit. The mesocarp is a thicker white spongy mass of tissue that surrounds the fleshy part of the fruit. The carpels which we eat are rich in vitamin C, and also contains a quantity of pectin; the extraction of both vitamin C and pectin from the fruit is an important industry in citrus-producing areas. The centre of the fruit consists of the much enlarged carpels of the ovary, each segment into which the fruit can be split being one carpel. The juice is produced by multitudes of hairs which grow out from the carpel wall; as the fruit develops these produce a large quantity of juice, flavoured by the essential oil characteristic of the species.

The banana fruit is also a berry, though it does not correspond with the popular idea of what a berry ought to be like. Cultivated bananas are triploid: that is, the nuclei of their cells contain three sets of chromosomes, the rod-like entities that carry the hereditary characters, instead of two sets as in normal (diploid) nuclei. For this reason normal fertilization of the flower in the cultivated banana is difficult, and the fruit develops without it. A cultivated banana contains no seeds, though you can see that the original ovary was composed of three carpels, and even make out the ovules as small brown objects in the centre of the fruit. The wild bananas are diploid (two sets of chromosomes, like most plants); their fruit develops normally, after fertilization, and contains seeds.

A banana has a pronounced curve, which is by no means accidental. The flowers of the banana are crowded together in long inflorescences which hang downwards. The flowers at the top of the inflorescence are male, those of an intermediate part are neutral, and the female flowers are at the bottom. As the fruits develop they curve away from gravity ('negative geotropism'), so that their tips point outwards. The fruit of the walnut (*Juglans regia*) and the coconut (*Cocos nucifera*) are both drupes, like the plum and the cherry, but before they are marketed the fleshy mesocarp is stripped off them, leaving only the hard endocarp surrounding the seed. In the endocarp of the coconut, as in many other drupes, the process of lignification, whereby the walls are hardened by the deposition inside them of the tough substance lignin (see Chapter 2), has spread beyond the vessels and fibres of the wood; practically everything is lignified, including the parenchyma cells of the cortex. The endocarp of the coconut is one of the hardest structures in the plant kingdom.

Inside the shell of the coconut we find the single seed. The seed coat is fused with the endocarp of the fruit. Within it is a small embryo and a large mass of endosperm. When young, the endosperm consists of a fairly thin outer portion enclosing a large cavity containing fluid, the 'milk' of the coconut. As the seed matures, more and more of the fluid becomes replaced by solid matter, though the milk never entirely disappears. The milk of a young coconut is a delicious drink, but to taste it one must go to the tropics where fresh young coconuts can be obtained. By the time a coconut reaches this country, what remains of the milk is very insipid.

Coconut milk is a good source of several unidentified substances that promote the growth of plants. When small pieces of the apical meristem of a plant—the tip of a stem or root where cell division is taking place actively—are placed in a solution containing all the substances known to be necessary for growth, cell division continues more or less unchecked. If, on the other hand, pieces of mature tissue such as portions of the phloem of the carrot root are similarly treated, cell division goes on very slowly or not at all. The addition of a small quantity of coconut milk to the culture solution will stimulate the cells into active division. Clearly the coconut milk must contain some substance or substances that act as growth promoters. An examination of coconut milk has confirmed the presence of at least six such substances.

Another instance of the growth-promoting effect of coconut milk is seen in the improvement of small immature embryos of the American thorn-apple (*Datura stramonium*) when placed in a culture fluid enriched with coconut milk and auxin. Without the coconut milk there is no growth. More mature embryos can do without the coconut milk; clearly at this stage they have either acquired the growth substances they need from the seed or, more probably, have become able to synthesize them for themselves.

Ocean dispersal theory (coconuts)

The means by which coconuts are dispersed has been a matter of controversy among botanists. The conventional story is that they are dispersed by floating in the sea, and the almost universal occurrence of the coconut palm on islands in the Pacific has lent credence to this fable. In most textbooks on elementary botany you will find the coconut quoted as an example of a fruit adapted to dispersal by ocean currents. This may well be true, but when one examines the evidence one sees signs of conflict: the stories of different observers do not tally with one another. It may be that the sea dispersal of the coconut is one of those facile theories that are believed because nobody has troubled to doubt them.

When Thor Heyerdahl and his companions drifted across the Pacific on the raft *Kon-Tiki*, the coconuts that they carried in baskets on the deck remained viable all the way from America to Polynesia; but other coconuts that were kept below deck, where they were in contact with the sea, were soon spoilt by the water or by the 'refuse collectors' that are constantly at work cleaning the oceans of anything organic that may be drifting around. On the other hand, a botanist in Honolulu floated coconuts in the water of Pearl Harbour for a hundred and ten days and found that they remained perfectly viable. During so many days, a coconut might well be carried three thousand miles by sea currents. It will be seen that these two observations directly contradict one another.

The only positive evidence of the ocean transport of the coconut comes from some volcanic islands, such as occur in the Krakatoa group, the site of the greatest eruption of historical times, in 1883. During 1928–30, a new island, Anak Krakatoa IV, was thrown up by volcanic action and remained to become colonized by anything that could cross the seas. About eighteen months after its appearance, the island was visited by Dr. van Leeuwen, who found the shore occupied by plants belonging to what is called the Barringtonia association. This plant community, often found along tropical sea shores, included *Barringtonia*, the screw pines (*Pandanus*), and the Indian almond (*Terminalia catalpa*). Among the other plants, Dr. van Leeuwen counted forty-one germinating coconuts. There seemed to be no doubt that the coconuts were sea-borne, for the island was still in a highly dangerous state, and no native canoes would have dared to approach it. Unfortunately we cannot tell whether the coconuts would have grown up into palms, for the island was destroyed by volcanic activity in 1932.

There are a few, but only a few, fairly reliable examples of coconut palms growing to maturity after being self sown. One is an area of coconut palms on the east coast of Trinidad, which are said to be the result of the wreck of a French schooner laden with coconuts.

Examining the evidence, it would appear reasonably certain that the coconut can travel by sea, sometimes for long distances, and establish itself where it makes a landfall. Yet such instances must be very rare. Apart from the difficulties and dangers of the sea crossing, the chances against a coconut seedling getting established after it and growing to maturity are great. There is the burning intensity of the tropic sun to be considered, as well as the difficulty of the seedling establishing itself among the welter of vegetation already growing. Even if, by a miracle, the seedling found a suitable place and began to grow in earnest, there would almost certainly be plenty of animals, such as wild pigs, to see that it came to nothing. We must regard the ocean transport of coconuts, therefore, as largely a myth.

How, then, does the coconut cross the sea? The answer is quite simple: it goes by canoe! For thousands of years primitive man has been using coconuts, and wherever man went in the tropics he took

164

the palm with him. Later—much later—the trading schooner helped. The inland spread of the coconut palm has been entirely due to man, and the seaward spread nearly so. The coconut is a fruit that is animal dispersed, the animal being man.

A strange associate of the coconut in the Pacific area is the robber crab (*Birgus latro*), which was observed by Darwin on the Cocos-Keeling Islands in 1836, during the voyage of the *Beagle*. This animal has the usual pincers on its front pair of legs, but they are exceptionally large and heavy. It also has a smaller, more slender pair on its last pair of legs. The crab tears the husk away from a coconut, fibre by fibre, always at the end where the three 'eye holes' are situated. When it has exposed the eye holes, it taps repeatedly on one of them until it has made a hole in the hard shell. It then inserts one of its hinder pincers into the shell and extracts the endosperm. This crab appears to feed solely on coconuts, and it is hard to imagine that it could have evolved its peculiar form without the coconut, to say nothing of the instinct that prompts it to tackle such an outwardly unpromising form of food. It is not only in the plant kingdom that we find natural selection producing a well-designed organism.

Mistletoe: method of dispersal

The mistletoe (*Viscum album*) has an interesting method of dispersal, which is in keeping with its parasitic growth on broad-leaved trees, especially the apple. The familiar white 'berries' of the mistletoe (they are not, in fact, berries, but false fruits in which the swollen receptacle of the flower is included) are attractive to birds, especially the mistle thrush. The fleshy part of the fruit contains a glutinous pulp which has, in fact, been used in making bird lime. When the bird eats the fruit it finds its beak entangled with this sticky substance, and in order to free itself from discomfort it strops its beak against the branch of the tree where it is perching. The mistletoe seed is rubbed off on to the bark, where it is held by the sticky remains of the fruit. In this way, the mistletoe is transferred from tree to tree.

Other dispersal mechanisms

A much stranger dispersal mechanism is seen in *Rafflesia arnoldi*, a parasite from the Malay Archipelago, which is said to be dispersed by elephants. *Rafflesia* is parasitic on the roots of forest trees. It has no aerial stem at all, living its life underground, but at flowering time it produces enormous flowers, sometimes a yard across, on the ground. The fruit is soft and fleshy (see page 141).

The problem for *Rafflesia* is how to get its seed to the roots of trees, buried beneath the soil. The flowers are trampled on by elephants, which squash the fruits underfoot. The seeds stick to the broad feet and are later pressed against the roots of trees by the weight of the elephants' tread. Only a beast as heavy as an elephant, with feet that sink deeply into the soft soil, could work this odd method of dispersal.

In Britain, when we speak of a fleshy fruit attracting animals it is birds we chiefly have in mind, though fruit-eating mammals, including man, may also come into the picture. In the tropics there are many more mammals to compete with the birds for the available fruit. Elephants, pigs, deer of various kinds, tapirs, monkeys and a host of others, including primitive man, join in the feast. Insects fight over the remnants left by the larger beasts.

A fruit is not ready for dispersal until it is ripe and the seeds it contains are fully formed. It would be a waste of effort for the plant to produce a heavy crop of fruit if all were to be eaten before the seed was ready. This is prevented, in succulent fruits, by the flavour, colour and smell appearing only when the seeds are mature. Most of us have experienced in our youth the painful consequences of raiding the orchard before the apples were ripe. Immature fruits usually contain distasteful substances, particularly organic acids such as malic acid, to rebuff the impatient. Only when the fruit is ripe do these acids disappear, to be replaced by sugar to give the fruit a sweet taste. At the same time, the colour of the fruit changes from green, inconspicuous against the foliage, to a brighter colour that will advertise its presence to all comers. It is not by chance that so many fruits are coloured red when ripe, for red is the colour most easily noticed against green by both birds and mammals.

Dry fruits may be dispersed by animals, usually by sticking to their coats, as in the various forms of burr fruits. These are usually provided with small hooks by which they can cling to fur or hide, or to the feathers and feet of birds. These hooks may be part of the fruit itself, as in the goosegrass or cleavers (*Galium aparine*) and the enchanter's nightshade (*Circaea lutetiana*) (Fig. 32), or to some other part of the flower that is included in the fruit to aid dispersal. In the wood avens (*Geum urbanum*) the style of the carpel persists instead of withering, and is hooked at the end (Fig. 33). The burr marigolds (*Bidens*) have a modified calyx; instead of consisting of a ring of hairs called the 'pappus', as it is in most of the family (Compositae) to which the burr marigolds belong, it has the form of two or three stiff bristles with an

Fig. 32.
Hooked fruit of the enchanter's nightshade

array of backwardly-directed barbs (Fig. 34). In the common agrimony (*Agrimonia eupatoria*) the calyx of the flower persists, covering the fruit, and this is equipped with hooks.

Burr fruits cling tightly to a rough surface, as anyone will know if he has ever tried to remove them from the coat of a dog or cat or, for that matter, from his own trouser-leg.

Many fruits are dispersed by the wind, and the number of adaptations that favour wind-dispersal is immense. The simplest wind mechanism consists of having the seeds small, light, and produced in large numbers; when the fruit opens, setting free the seeds (and in some cases projecting them explosively from confinement), the seeds will be carried away by the breeze. This is the method in the foxglove

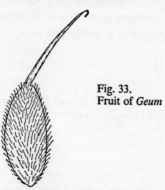

Fig. 33.
Fruit of *Geum*

(*Digitalis purpurea*), and it reaches perfection in many of the orchids, where the seeds are so minute that they form a powder.

Sometimes the seeds themselves are provided with plumes of hairs to catch the wind, as in the rosebay willowherb (*Chamaenerion angustifolium*). The fruit opens while still attached to the plant, exposing the tufts of hairs (Fig. 35), and on a windy day the air is full of floating seeds.

Winged fruits are common, as in the elm, the ash and the sycamore; the wing is usually formed by a flat expansion of the pericarp.

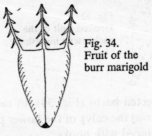

Fig. 34.
Fruit of the
burr marigold

The form of the wing varies from plant to plant. In the elm it is oval (Fig. 36), while in the ash it is long and somewhat twisted (Fig. 37). The sycamore has a two-seeded fruit that splits in half when ripe, though the two halves do not separate; the seeds remain attached, with the stalk between them and the two long, flat wings projecting on either side, rather like the propeller of an aeroplane (Fig. 38).

It is in the daisy family (Compositae) that we see the most perfect adaptations to aerial transport. The 'flower' of a daisy, sunflower or single chrysanthemum is not really a single flower, but a collection of flowers. If we carefully pull out one of the petals from a sunflower, we find that it is in fact one very small flower. If we then examine the 'eye' (disc) we find that it too is composed of a multitude of tiny

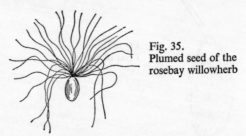

Fig. 35.
Plumed seed of the
rosebay willowherb

flowers, though they lack the long, strap-shaped petals that glorify the outer flowers. In a large flower head there may be hundreds of these minute florets, as they are called. The idea of having the florets grouped together in a flower head or 'capitulum' is to make the whole structure conspicuous for pollination while having the individual flowers small.

Some of the Compositae have both disc florets, without the strap-shaped petal, and ray florets with the conspicuous corolla, which is really composed of five petals fused together, though there is often no outward sign of it. In others the florets are all 'ligulate'—that is, with strap-shaped petals. An example is the common dandelion.

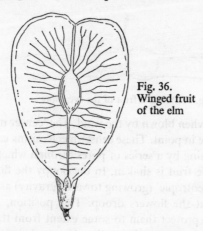

Fig. 36.
Winged fruit
of the elm

The individual floret in the Compositae is much simplified. Though formed from two carpels the ovary has only one cavity and contains a single seed. The calyx is usually represented by a ring of hairs, the pappus, and this serves as a parachute that enables the small fruit to hang in the air, sometimes for hours, for every small

Fig. 37.
Winged fruits
of the ash

breath of wind will prevent it from settling. A composite fruit may travel miles in the air before it finally comes to rest.

In some composites, such as the dandelion (Fig. 39) and the goat's beard or Jack-go-to-bed-by-noon (*Tragopogon pratensis*), there is a further refinement. As the fruit develops, the pappus becomes separated from it by a long stalk. You can see this quite clearly in a dandelion 'clock'. In the goat's-beard the hairs of the pappus are feathery with a multitude of small branches, which makes them even more efficient as a parachute.

Some plants scatter their seed from the fruit, aided by the move-

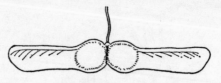

Fig. 38. Winged fruit of the sycamore

ment of their stems when blown by the wind. The 'censer mechanism' of the poppy is a case in point. These censer mechanisms usually have their ripe fruits opening by a series of pores through which the seeds can fall out when the fruit is shaken. In the poppy the flower stalks become 'positively geotropic' (growing towards gravity) as the flower buds appear, so that the flowers droop. This position, with calyx uppermost, helps to protect them to some extent from the weather. As the fruits form, however, the stalks change their response to gravity, becoming 'negatively geotropic' and growing upwards. The flower stalk is somewhat elastic, and makes sharp, whipping movements when the wind is blowing, and these cause the seeds to be thrown out of the fruit. Since the openings are at the top of the fruit, this ensures that the seeds will be flung out, instead of just falling, so

Fig. 39.
Fruit of the
dandelion

that some of them at least will come down at a distance from the parent plant, where they will have a chance to grow up without being overshadowed. This, incidentally, is one of the objects of most seed dispersal mechanisms.

The flower stalk of the poppy changes its reaction to gravity at the last moment in order to assist dispersal of its seeds. In some plants the change in reaction is not to gravity, but to light. The ivy-leaved toadflax (*Cymbalaria muralis*) is a delightful little plant that has become quite common in Britain since its first appearance in 1640; it grows in cracks in old walls. The flower stalks are at first 'positively phototropic' (growing towards light), so that the lilac flowers are held out for pollination by bees, but as the fruits begin to set a change in the response to light takes place. The flower stalks become 'negatively phototropic', shunning the light and growing into crevices in the wall. Here the fruits deposit their seeds.

In some plants we see a more sophisticated version of the catapult action of the poppy flower stalk, for the fruit is explosive, ejecting its seeds to a distance by a sudden boisterous eruption. In the common gorse (*Ulex europaeus*) the fruit pods, when ripe, are in a state of tension because the different layers of the pericarp dry at different

Fig. 40.
An African lucky bean,
with its elaiosome

rates. The slightest disturbance will cause a ripe fruit to burst open, scattering its seeds in all directions. On a summer day among gorse bushes, if one can get far enough from the clangour of the internal combustion engine or the banshee wail of jet aircraft, it is possible to hear the popping of gorse pods as they eject their seed.

The gorse does not rely solely on its explosive mechanism to secure the dispersal of its seeds, for the seed itself is provided with another device which is probably more effective than the ebullient pods. Each seed has a fleshy, orange appendage which is attractive to ants. The ants can be seen biting and tearing at it, and dragging the seed along as they do so. It is said that this is the reason why gorse tends to occur along footpaths, for the ants follow the tracks, dragging the gorse seeds with them.

Organs that attract ants are called 'elaiosomes', and they are quite common in the plant world, occurring either on the seed or the fruit. Usually they are impregnated with oil, which ants love. The nutmeg

171

and the African lucky bean (Fig. 40) are well-known examples, and nearer home elaiosomes are found in the fumitories (*Fumaria*), bugle (*Ajuga reptans*), fingered sedge (*Carex digitata*) and many others. Plants of the onion genus (*Allium*) and many other monocotyledons have thin outer seed coats that are impregnated with oil. Seed dispersal by ants appears to be particularly important in woodland plants. Ants, after all, are primarily woodland dwellers.

Some aquatic plants have fruits or seeds that are dispersed by water, though here it must be remembered that a water plant may use some other agency; many, for example, are dispersed by birds. Fruits that are intended to float in water usually have some adaptation to render them buoyant. The fruit wall of the water plantain (*Alisma plantago-aquatica*) is loose and spongy, enclosing air that assists floating, and the seeds of the yellow flag (*Iris pseudacorus*) likewise have seed coats with corky, air-containing tissues. We have seen that the coconut, the fruit that is often quoted as being the greatest ocean traveller of all, is not in fact dispersed by sea; but a number of maritime plants are, including the tropical *Barringtonia* that is such a common feature of the shore in certain places. The fruit of *Barringtonia* has aerated tissues that help it to float, but fruits may be dispersed by sea water without these aids. Along our own coasts we find the sea kale (*Crambe maritima*), the sea rocket (*Cakile maritima*), the sea beet (*Beta vulgaris* var. *maritima*), and species of glasswort (*Salicornia*) being carried from place to place by the sea.

The mangroves are often dispersed by ocean currents, but with many of them it is not fruits or seeds but seedlings that are the voyagers. The mangrove is not a single plant or even a genus of plants; its peculiarities are dealt with on pages 59, 117 and 192. It is a general name given to plants of a particular habit and habitat. Many mangroves are 'viviparous': their seeds germinate while still carried in the fruit attached to the parent plant. The young root of the seedling grows straight downwards, directed by gravity, and when the young seedling at last is dropped from the fruit, it plunges down like a dart, root first, and plants itself in the mud of the mangrove swamp. If it should fall in water instead of mud, it may travel a long distance before it finds itself ashore, apparently without harm.

Some fruits and seeds have no special means of dispersal, and yet one generation follows another. The rayless mayweed (*Matricaria matricarioides*) is a common weed found by waysides. Its fruit is small and wingless, and the hairy pappus characteristic of many

Compositae is absent or vestigial, yet this plant is spreading rapidly throughout the British Isles. It is thought that it owes its wide dispersal largely to fruits embedded in the mud caked on motor car tyres. Other weeds may well be spread in the same way.

Not only are seeds distributed in space, but some of the most successful weeds are also distributed in time. In other words, their seeds do not all germinate in the same year. There is an old farmers' adage that 'one year's weed means seven years' seed', and this piece of bucolic wisdom is nothing if not conservative: about forty years, for instance, instead of seven, would be nearer the mark in the case of charlock. There are records of this pernicious weed springing up from soil that had been under woodland for at least forty years, and the same is true of the curled dock (*Rumex crispus*), the great plantain (*Plantago major*), and certain other weeds of waste places and arable land.

I have dealt earlier with the question of longevity of seeds. The figures that I quoted on page 119 were for seeds stored in more or less dry conditions. In the soil they are not dry, and it is much more difficult to see how dormancy is prolonged under natural conditions as opposed to a dry herbarium sheet. Little is really known about seed dormancy under natural conditions. In some cases a long period of dormancy may no doubt be accounted for by the hardness of the seed coat, but this cannot explain why a seed should remain dormant in the soil for a period of years. In such a time the seed coat must surely have softened enough to permit germination.

Some work was done on the germination of seeds of the black mustard (*Sinapis alba*) that may have a bearing on prolonged dormancy in the soil. It was found that seeds that were saturated with water failed to germinate if there was more than a certain concentration of carbon dioxide in the surrounding air. Carbon dioxide is liberated by the decay of organic matter in the soil, and it may well be that this is the factor that is largely responsible for much of the delayed germination that one encounters here and there. Further work is needed, however, before we are justified in forming a theory about this.

The advantage to a plant of not having all its seed sprouting at the same time is that by the spreading of germination over a longer period there is opportunity for profiting from a variety of different climatic situations. If all the seed germinated together it might be that unfavourable conditions, such as a drought following germination, would cause a loss of most or all of the seedlings.

173

10 · Plant survival

In a tropical rain forest plants can grow all the year round, for the constant high temperature and humidity render a dormant period unnecessary. In most other parts of the world there is a period in each year that is unfavourable for plant life. In Britain that period is the winter, when frozen soil and icy winds put a stop to the growth of all but a hardy minority of plants. In other countries it may not be the cold so much as the dry season of the year that is unfavourable; this is the case, for instance, in the regions where the monsoon forest is the climax of vegetation, such as some parts of India. There the burning heat of the sun during the annual dry season is as much a bar to plant life as the winter frosts of Britain.

Some plants can escape the unfavourable period altogether by surviving only as seeds. These are the annual plants. They live their lives during the favourable period and then die completely, leaving only their seeds behind them to carry on the species. Sometimes there may be several generations in the course of a single season, as in the chickweed that is so common everywhere on cultivated ground. This little plant takes but a few weeks to grow, come to maturity, and set seed.

In some respects this may be the ideal way of coping with the problem of the unfavourable season, for nothing is better adapted for survival under trying conditions than a seed. Its thick seed coat, and the state of dormancy into which it is plunged, usually ensure that no freaks of climate can cause it serious harm. But the annual habit still carries with it one serious disadvantage. Since all the growing has to be done in one year, there is a limit to the size that the plant can attain. Most annual plants are small.

A plant that survives from year to year is called a perennial, and in order to survive the harsh season it must have some means of keeping at least part of itself alive to grow again next year. This it

174

does by means of winter buds, or, to be more precise, perennating buds, for the unfavourable period may not necessarily be the winter.

Woody perennials—trees and shrubs—carry their buds on their branches. The plant may lose its leaves during winter, but the buds remain to burst into life in spring. Not all perennials are woody, however. Herbaceous perennials, such as the delphinium, lose their soft, herbaceous stems when winter comes, but the rootstock remains alive, though dormant, and on this are the perennating buds. These are protected from winter frosts by the litter of dead leaves and the soil around them. When the spring sunshine warms the ground again the buds spring into life, and new aerial shoots are formed.

The fact that plants can be divided into a number of 'life forms' was first recognized by the Danish botanist Raunkiaer, more than sixty years ago, and his ideas are still valid. The various life forms are arranged according to the position of their perennating buds (in the climate of Britain these are synonymous with 'winter buds'). Thus, the ones that carry their winter buds freely exposed in the air, as do trees and most shrubs in this country, are called 'phanerophytes'; those that bury them below ground, like the crocus, are 'geophytes', and so on. There are seven principal life forms recognized today.

Phanerophytes (a life form)

The phanerophytes include all the trees and nearly all the shrubs. Their buds are carried in the air, well above ground level, and so are exposed to the worst that the winter season can do to them. Perhaps this is why the phanerophytes are predominantly plants of warm climates, where frost plays little or no part in the hazards that they have to face. This does not mean that some phanerophytes have not become adapted to colder places, for the northern coniferous forest, the vegetation type that stretches from north America, right across northern Europe, into Siberia, is proof that some phanerophytes can withstand cold. But the number of species, as opposed to individuals, of these northern phanerophytes is very small when compared with the number found in the tropics. It is to the equatorial regions of the world that we must go if we wish to see trees in their full glory of luxuriance. In a tropical rain forest of quite small dimensions the number of different species of trees may run into hundreds.

In such a tropical rain forest the buds may be quite unprotected by scales, for with the absence of any unfavourable period there is nothing to prevent them from developing straight away. In regions where there is a pronounced dry period we find various methods of bud protection becoming increasingly common. The outer leaves may be covered with hairs that trap water vapour between them and so check its evaporation from the buds. The buds may be covered with sticky mucilage or resin, which serves the same purpose. Still further protection is obtained in some plants by having the buds protected by the stipules of the leaves—the leaf-like outgrowths at the base of the leaf stalk that are found in many leaves.

As we come to colder parts of the world where the winter may be long and severe, we usually find the buds protected by a covering of scales; this may be seen, for instance, in most of the phanerophytes found in Britain. Sometimes the protection given by the bud scales may be reinforced by hairs on the scale, or by a gummy covering of mucilage, as in the well-known 'sticky buds' of the horse chestnut.

Another adaptation of phanerophytes to unfavourable conditions is autumn leaf-fall. In the tropical rain forest evergreen trees prevail; most of our own trees and shrubs are deciduous—they lose their leaves on the approach of winter. This is to guard against too much loss of water through transpiration at a time when the coldness of the soil makes it difficult for the roots to absorb enough for the needs of the plant. More will be said about the autumn leaf-fall in Chapter 11, in connection with xeromorphic adaptations.

The farther the phanerophytes penetrate into high latitudes the lower in stature they tend to become. The same is true with high altitudes. We can see this in Britain with some of our trees and shrubs. The dwarf birch (*Betula nana*), for instance, is not found south of Northumberland, and occurs chiefly on moors at an elevation of from eight hundred to two thousand eight hundred feet; it extends north-wards to Iceland. It is a shrub no more than three feet high, and its branches tend to spread out close to the ground. The nearer the ground, the better protected are the winter buds.

Chamaephytes (a life form)

Phanerophytes such as the dwarf birch bring us naturally to the second life form, the 'chamaephyte'. Here the perennating buds are carried at soil level, or just above it. The aerial shoots that bear the flowers project freely up into the air, but they die back in autumn, leaving the portions that bear next year's buds on or near the ground. In a cold winter they may be protected by a covering of snow, which serves to prevent desiccation and at the same time keeps out frost. In warmer regions, where drying rather than freezing is the danger, they are usually covered by a litter of dead plant remains. Among common British chamaephytes are the greater stitchwort (*Stellaria holostea*), the creeping Jenny (*Lysimachia nummularia*), and the common speedwell (*Veronica officinalis*).

Hemicryptophytes (a life form)

Where there is a pronounced unfavourable season, and particularly where the unfavourable season is cold, the buds need better protection than can be provided by the chamaephyte. The next stage in bud protection is seen in the 'hemicryptophyte'. Here the winter buds are formed in the surface layers of the soil itself, where they are protected not only by the soil surrounding them but also by the withered remains of the aerial shoot that dies when the cold weather begins.

Hemicryptophyte sub-categories

The hemicryptophyte shows an advance in bud protection over the chamaephyte, and it is not surprising that most of our British herbs, and about half the plants of central Europe, belong to this type. As in the other life forms, the hemicryptophytes are subdivided into categories that show increasing stages of bud protection. Starting with 'protohemicryptophytes', with flowering shoots bearing leaves all the way up to the inflorescence, there is progress through the 'semi-rosette' plant, with the lower leaves distinctly larger than the upper ones, to the 'rosette' plant in which the erect stem bears only the flowers, the leaves being confined to a rosette at the base of the stem, as in the familiar dandelion.

The winter buds of hemicryptophytes are kept in the surface layers of the soil, and do not become deeply buried. Their protective covering is obtained by various means, among which the deposition of soil by earthworms plays quite an important part. The roots of many hemicryptophytes are contractile, dragging the plant down into the soil if it should, in the course of time, become uncovered.

Sometimes the underground shoot system becomes buried too deeply, owing to the activity of moles or some other cause. When this happens, the plant produces its perennating buds higher up the shoot, so that they again lie just below the surface of the soil.

Many hemicryptophytes have stolons—creeping stems that root at intervals, serving for vegetative reproduction. These may run above ground, or just beneath the surface. The strawberry is an example of a hemicryptophyte with stolons that run above ground, while the butterbur (*Petasites hybridus*) has underground stolons. Some plants, like the stinging nettle (*Urtica dioica*) and the hedge woundwort (*Stachys sylvatica*), have both types of stolon.

Rosette plants

The form of rosette plants such as the dandelion seems to be, to some extent, determined by light. The rosette habit is brought about by a shortening of the 'internodes'—the spaces between the leaves—of the stem. While the stem is growing underground it has normal internodes, but as soon as it comes into the light the internodes shorten, so that the leaves become crowded together. If one piles soil on top of a dandelion plant just as it is about to emerge from the soil, the internodes lengthen until the tip of the stem again emerges into the light; then they shorten and the plant assumes its normal rosette habit. The flowering shoot reacts differently. Its growth is not checked by light, and so it rises up into the air. It does not produce leaves.

Geophytes (a life form)

From the hemicryptophytes that have their winter buds in the soil surface, it is an easy step to the next life form, the 'geophytes'. In these the winter buds are buried in the soil during the winter season, the shoot only growing up and emerging from cover when the weather is favourable for growth. All the bulbs and corms grown by the gardener, such as the daffodil, tulip, hyacinth, and crocus, belong to this category.

The geophytes have the most perfect protection for their perennating buds that any plant can devise. Buried in the soil, they need fear neither desiccation nor frost. They are particularly well adapted for life in regions where there is a prolonged dry period, and it is natural that one should find them the predominant life form of dry steppes. Yet many geophytes are adapted to live in districts where the growing season may be long but the winter is severe. The myriads of crocuses that delight the eye in the Swiss Alps in spring are instances of this.

Where geophytes live in a region where the growing season is short, as in the steppes, it is essential that they should make rapid progress when they start. The aerial shoot bearing leaves and flowers *must* develop quickly. In a geophyte, therefore, there must be some means of storing food underground, where it will be immediately available for the growth of the aerial part of the plant when the favourable season comes round. Many geophytes are provided with bulbs or corms in which food is stored until it is needed. Others are provided with thick underground stems (rhizomes) which serve the same purpose.

Helophytes (a life form)

Plants that grow in marshes, where the soil is saturated with water, and which place their winter buds below the surface of the soil, are called 'helophytes'. The reed maces (*Typha*) are an example. Some helophytes such as the arrowhead (*Sagittaria sagittifolia*), may grow in shallow water, leaving their perennating buds buried in the mud at the bottom.

Hydrophytes (a life form)

Helophytes such as these merge into the next class which are the 'hydrophytes', with buds covered and protected by water during the unfavourable season. Many of our familiar water plants belong to this group, including the white and yellow water lilies (*Nymphaea alba* and *Nuphar lutea*), where the winter buds are found on rhizomes at the bottom of the water in which the plants grow. Some hydrophytes have no rhizomes. Their winter buds become detached from the plant and sink to the bottom of the water, there to await the favourable season. The grassy pondweed (*Potamogeton obtusifolius*) is a hydrophyte that behaves in this way.

Therophytes (a life form)

The seventh and last life form consists of those plants that die completely at the end of the growing season, trusting to their seeds for survival. These are the 'therophytes', and we have already dealt with them. In some respects they may be regarded as the most perfectly protected of all, for there is nothing so resistant to adverse conditions as a seed.

Winter buds, like seeds, usually enter a state of dormancy from which they cannot normally be aroused until it is time for them to start growing. This is clearly a biological advantage to the plant; it would be useless to protect the winter buds if they were to begin growth prematurely, exposing the tender young shoots to cold or drought. Dormancy of buds is a general phenomenon, absent only in particularly favourable environments such as the tropical rain forest, but it is especially noticeable in phanerophytes and cryptophytes that grow in temperate climates, where the winters are cold.

Dormancy of buds

In the climate that prevails in northern and central Europe, including the British Isles, exposure to cold appears to be the factor that normally operates in Nature to break the dormancy of buds. In the phanerophytes native to Britain, the winter buds need exposure for a period to a temperature below 9°C before they will begin to grow. The length of the exposure that is needed varies with different plants; for some as short a period as two hundred hours is sufficient to break dormancy, while for others an exposure to cold of several thousand hours is necessary.

Dormancy can be broken artificially by various means, such as treatment with acids or gibberellin (a substance produced by the fungus *Fusarium moniliforme*), exposure to the vapour of ether or hydrocyanic (prussic) acid, the injection of the bud with alcohol, or immersion of the twigs in lukewarm water for a few hours followed by placing them in a warm greenhouse. Nobody knows why these treatments are effective.

The internal factors that produce dormancy have not as yet been elucidated with certainty. One might expect auxin (see Chapter 4) to be concerned, since auxin inhibits the further development of buds that arise behind the growing apex of the stem, but apparently it is not a factor in dormancy. In some cases it seems as if an inhibitory substance is produced in dormant buds. If the buds and leaves of the sycamore (*Acer pseudoplatanus*) are extracted with alcohol, a substance is obtained which slows down the growth of wheat coleoptiles and certain other parts of plants. It has been shown that this substance, which is at present unidentified, is contained in the buds in greatest quantity during early winter, and least when the plant is growing actively; at no time is it entirely absent.

An important feature of this rather mysterious inhibitory substance is that it appears to be produced at least partly as a result of
illumination of the plant by short days. The day-length (i.e., the
period during which a plant is illuminated during each twenty-four
hours) is of fundamental importance to plants, as we shall see in the
next chapter, and it would not be at all surprising if it were implicated
in the onset of dormancy in buds. It was found that more of the
inhibitor was produced in the leaves of sycamore plants which were
artificially given short days than in similar leaves which received
normal illumination, and it is suggested that, in Nature, the shorter
days that herald the autumn may be responsible for the increase in
the inhibitor, causing the buds to become dormant.

The importance of short days in producing dormancy has been
noticed with other plants, which showed delayed dormancy if they
were given extra illumination as the days became shorter with the
passing of the year. The tulip tree (*Lyriodendron tulipifera*) has been
shown to maintain its growth right through the winter if given a long
day by extra illumination, combined with a suitable temperature.
Moreover, the extra light need not be very strong, which disposes of
any idea that the delay in dormancy might be due to increased photosynthesis.

Whether other plants besides the sycamore produce an inhibitor
which induces dormancy in buds is not yet known. We are also
ignorant of the effect of short days in inducing dormancy, and can
speak only for the few plants that have been investigated. There is
room for a great deal of research on the dormancy of buds, and the
results would certainly be rewarding in scientific interest and, possibly,
in commercial application.

11 · Adaptation to habitat

When plants first came out of the sea and began to live on land, they had to adapt themselves to a completely different environment, and we saw in Chapter 1 how some of this adaptation was accomplished. A root system was developed to anchor them in place and to absorb water and nourishing minerals from such soil as there was, and conducting tissues were formed to convey the water and minerals to the parts where they were most needed. Land plants were not born in a day; they needed millions of years of constant evolution before they were really suited to terrestrial life.

This process of adaptation has gone on ever since, and is still going on. Plants cannot afford to remain static, for that way lies extinction. For one thing, the environment changes over the years, and a plant that is nicely adapted to a particular environment may find its surroundings altering. The climate may become warmer or colder, wetter or dryer. Any plant that cannot adapt itself to the changing environment must either perish or move somewhere else.

When the late Palaeozoic era was succeeded by the early Mesozoic, the climate changed dramatically. The swamp forests of the Carboniferous gave way, during Permian times, to the deserts of the Triassic. The effect on plant life was cataclysmic. Many species gave up the struggle and ceased to exist, while those that were left found the regions they could inhabit dwindling around them. At the same time, competition increased from new species that were better designed to meet the new conditions.

The change in environment was both a challenge and an opportunity. The old plants failed in the main to meet the demands that Nature made on them. The changes that spelt extinction for some species meant a stimulus to the evolution of others. As always, a time of risk was also a time of opportunity.

A more recent catastrophe occurred during Pleistocene times, when the climate of Britain became cold without precedent, and the ice spread from the North Pole almost to what is now the Sussex coast. Plants fled before the advancing ice, only a few hardy species surviving, perhaps, in unfrozen pockets called nunataks. When the climate again warmed up and the ice retreated, only a proportion of the plants came back; many never made the return journey, and are lost to our flora for ever.

It is believed that there were four great ice ages during the Pleistocene period, separated by three 'interglacials'. The first ice age began about six hundred thousand years ago, and it is about twenty thousand years since the ice finally began to recede. Whether we are at present living in an interglacial, with more ice to come, nobody knows.

Although conditions on the earth are at present comparatively stable (in the physical sense), competition in Nature is still severe. There is an unending struggle of plants against their environment—and by 'environment' I mean all the factors of climate, physiography and soil that together make up the place where they grow, as well as the effect of the other neighbouring organisms, animal and plant, and of their own fellows of the same species that are competing with them for space and food. A plant needs to be extremely well adapted to its environment if it is to stand a chance of continuing to live.

Some plants seem to be able to stand four-square to the vicissitudes that beset them. The common dandelion, for instance, is one of Nature's tough guys. It can grow almost anywhere where there is soil, and even where there is seemingly none. It can barge its way into what is an apparently closed community, and having got there it is prepared to fight anyone or anything before it is evicted. Even when bodily pulled up, it will grow again from some tiny portion of its root system that has been overlooked by the would-be destroyer.

Not all plants have the resilience of the dandelion. There are some that avoid battle with other species, and prefer to retire to some out-of-the-way corner where competition is less severe, and they may have a chance of living out their lives undisturbed.

Pioneering plants

Such plants are not the vegetable kingdom's equivalent of cowards and snobs. Quite the contrary. They are the pioneers, and may properly be compared with adventurous folk who brave the unknown

183

and turn away from places that are comfortable but overcrowded. These plants have managed to capitalize on some quirk of structure or physiology that enables them to live where others would die. They have seized on some particularly unpromising area of the environment, where most plants cannot grow, and thriven there. In this way, the cacti have invaded the inhospitable deserts of the world, the edelweiss (*Leontopodium alpinum*) has made the high Alps its own, and the eel-grass has even put the clock back three hundred million years and established itself in the sea.

Xeromorphic characters

A plant that lives in a particularly dry situation is called a 'xerophyte', and the features that enable it to do so by cutting down undue loss of moisture are called 'xeromorphic characters'. Xerophytes and xeromorphy do not necessarily go together. A plant that lives in a dry situation, but manages, by rapid growth, to complete its life history during the brief wet season of the year, might fairly be called a xerophyte, though it would probably show no xeromorphic characters, because it would not need them. On the other hand, most saltmarsh plants show well-developed xeromorphy, although their roots are constantly bathed in water, and the aerial parts may be submerged at every high tide. The reason for this is that the water is salt, so that the roots have difficulty in absorbing it, owing to its high osmotic pressure (or low diffusion pressure). Such plants suffer continually from what is known as 'physiological drought'; with them it is a case of 'water, water everywhere, nor any drop to drink'.

Xeromorphic plants may show various characteristics that tend to cut down water loss. Their leaves may be thick and leathery, their cuticles being much thickened and often reinforced with extra wax. The development of hairs on the leaves is a frequent xeromorphic character. The presence of hairs, in which a layer of air is trapped, prevents turbulence from approaching the stomatal openings in windy weather, thus slowing down the evaporation of water vapour from the stomata. The sinking of the stomata in pits below the general leaf surface is another way of cutting down transpiration; water vapour collects in the funnels above the stomata, and is not readily blown away by the wind. Xeromorphic leaves often have a layer of thick-walled cells (the hypodermis) just below the epidermis. This probably serves a double function, checking evaporation and at the same time strengthening the leaf, so that it does not shrink easily on losing water.

Extreme xeromorphs may go far beyond these simple precautions against loss of water. The whole morphology of the leaf may be adapted towards this one necessary function. In the leaf of the oleander (*Nerium oleander*), for instance, the distribution of the stomata is restricted to a series of hollows, which themselves are guarded by hairs. In the ling (*Calluna vulgaris*) the stomata are confined to a groove (the stomatal chamber) running along the lower side of the thick, fleshy leaf.

Many grasses have leaves that can roll themselves up, the stomata being on the inside of the roll. The ability to do this depends on the presence of groups of cells called 'hinge cells', which occur on the inner surface of the rolled leaf. A particularly good example of this is the marram grass (*Ammophila arenaria*), the large, rolled leaves of which are a common feature of almost any sand dune. In the marram grass, the outer surface of the rolled leaf is relatively smooth, and bounded by a thick cuticle; it bears no stomata. The inner surface is thrown into a number of folds, at the bottom of each of which is a group of hinge cells. At the side of every group of hinge cells, within the fold of the leaf, is a patch of photosynthetic tissue, the cells containing a rich supply of chloroplasts. The stomata are confined to this region. As additional protection against excessive transpiration, the outer edges of the grooves are armed with many stiff hairs.

An advantage of this arrangement is that transpiration can be adjusted to the needs of the plant. When the leaf has sufficient moisture, the hinge cells become turgid and swell; this has the effect of opening out the leaf, so that carbon dioxide can enter the stomata freely. Should this result in undue loss of water, the hinge cells lose their turgor and collapse, so that the leaf rolls up.

The leaves of xeromorphic plants are often small, and may be reduced to scales, the photosynthetic function being taken over by the stem. An extreme case of this is seen in the she-oaks (*Casuarina*), of Australia. Here the leaves are small scales, and the twigs are deeply cut into a number of grooves around their circumferences. The green photosynthetic tissue is at the bottoms of the grooves, which are protected by hairs. All the work of photosynthesis is done by the green twigs.

Not only do the stems of many xerophytes take over the functions of leaves: in many cases they actually look like leaves. The butcher's broom (*Ruscus aculeatus*), a member of the lily family that grows in dry woods and among rocks in the south of England, has what appear to be small, leathery, pointed leaves (Fig. 41). Close examination reveals that the 'leaves' are branches of the stem. In the centre of the upper surface of each 'leaf' is a tiny scale leaf, and, in the flowering season the flowers are borne on these peculiar flat branches.

Stems that masquerade as leaves in this way are known as 'cladodes'. Sometimes the stalk (petiole) of a leaf takes on a leaf-like form, the blade of the leaf being lost. We see this in many species of *Acacia*. Such structures are called 'phyllodes' (Fig. 42).

It may seem strange that a plant, having lost its leaves in the interests of water economy, should develop leaf-like stems or petioles,

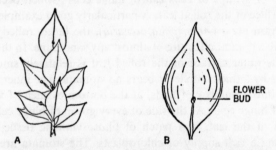

Fig. 41. **A**, part of the shoot of the butcher's broom, showing the leaf-like cladodes. **B**, single cladode, showing the flower bud on its upper surface

apparently throwing away the advantage that it has gained. Stems and petioles, however, commonly have more strengthening tissue than leaves. In the event of water shortage, they are better able to resist shrinkage than leaves, and so are less vulnerable.

Many xeromorphs bear spines which are modified leaves or branches. In the sloe (*Prunus spinosa*) and the hawthorn (*Crataegus monogyna*) the spines are modified branches, while in the barberry (*Berberis vulgaris*) they are modified leaves. The gorses (*Ulex*) have both branch and leaf spines. The seedling of the gorse does not bear spines until the stem has reached a certain height; it is thought that this is because the air near to the surface of the ground is comparatively moist, so that xeromorphic adaptations are not needed so low down on the plant.

Spines serve a useful function in keeping off browsing animals.

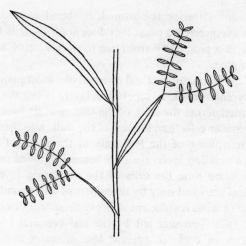

Fig. 42. Part of a shoot of *Acacia*
showing the transition from
leaves to phyllodes

It is an interesting thought that a characteristic acquired as a xeromorphic adaptation may be retained for quite another reason. This would be quite in keeping with the economical habits of plants.

The shedding of leaves during the winter by deciduous trees is a precaution against water loss. During the cold winter months the roots find it difficult to absorb water from the soil, and if trees such as the oak and beech kept their leaves all the year round, their water balance would be in danger during the cold weather. By shedding their leaves in the autumn this danger is avoided. Of course, it means that photosynthesis will cease when the leaves fall, but as growth is slowed down to vanishing point during the winter, this does not particularly matter. Evergreen trees and shrubs that keep their leaves all the year round usually have xeromorphic leaves. Consider, for instance, the pine, with its long, narrow, leathery needles. Examination of a cross section of a pine leaf shows the usual xeromorphic characters: thick cuticle, sunken stomata, thick-walled hypodermis, and a rounded shape that cuts down the ratio of surface to volume.

Autumn leaf-fall

Although a pine is spoken of as an evergreen, that does not mean that its leaves last for ever; a walk through a pinewood, with the thick

187

carpet of needles covering the ground, is abundant evidence to the contrary. An evergreen is a plant that does not lose *all* its leaves *every* year. The life of a pine leaf is from two to twenty years, according to the species of pine.

Autumn leaf-fall is not a casual process; like most plant functions, it is carefully engineered. A separating layer, called the 'abscission layer', is formed across the stalk of the leaf, near its base. This layer usually has smaller cells than the rest of the stalk, and there may even be a slight narrowing of the leaf stalk at the point of separation. Beneath the abscission layer, the cells become lignified or suberized, while at the same time the cells of the abscission layer separate, leaving the leaf attached only by its epidermis and vascular bundles. The least puff of wind is sufficient to break these tenuous connexions, and the leaf falls. The scar left by the leaf becomes covered by a protective layer of cork, to prevent loss of water and, still more important, to stop the entry of fungal spores.

Before the final death of the leaf, certain chemical changes go on inside it. The chlorophyll decomposes, so that the yellow colour of the xanthophyll can be seen; at the same time, anthocyanin pigments are developed in many leaves. The yellow and red autumn colours are a result of these processes. Decomposition products of chlorophyll may also take a hand in the colouring.

Succulents

The xeromorphic adaptations that we have so far discussed are all aimed at reducing transpiration, but they by no means exhaust the powers shown by plants of adapting themselves to dry situations. There is another line of approach in the development of water storage tissues. These are seen in succulent plants, which include some of the most extreme xerophytes such as the desert-living cacti, or, nearer home, by the stonecrops (*Sedum*), and the butcher's broom.

In many succulents there is an area in the leaf or stem occupied by large, colourless, thin-walled cells with the function of storing up water for future use. In the stonecrops, the prickly saltwort (*Salsola kali*), and the seablite (*Suaeda maritima*) the water storage tissue is in the fleshy leaves, while in *Cactus*, in which the leaves are reduced to spines, it is in the succulent stems.

A succulent plant can withstand a considerable degree of desiccation. As the water in the storage tissue is used up the plant shrinks, but apparently the shrinkage does not cause any harm, for the plant readily regains its size when given water.

Leaves of water plants

Water plants have to face conditions that are the opposite to the rigours endured by xerophytes. Their roots are always bathed in water, and their shoots also are submerged to a greater or less extent. Water is in many ways a favourable environment for plants, but it has its own special problems.

Conditions for plant life in water are far more uniform all over the world than conditions on land. This is reflected in the wider distribution of many water plants, compared with land plants. Water has a high specific heat: it is slow to warm up, and equally slow to cool down. Consequently, a water plant is not subjected to the violent and often rapid changes of temperature that occur on land. This favours the vegetative growth of the plant.

The conditions of nutrition of a water plant are also different from those of a land plant. Carbon dioxide is more soluble in water than oxygen, and, as a plant needs more carbon dioxide for photosynthesis than oxygen for respiration, this is favourable to vegetative growth. Water plants often show a luxuriance of growth, combined with prolific vegetative reproduction that is seldom seen on land, at least in temperate regions.

Water plants absorb not only carbon dioxide but also mineral salts over the whole of their submerged surfaces. Correlated with this, the cuticle of the submerged stems and leaves is thin, or absent altogether. Submerged leaves are deeply divided, often looking like bunches of hairs, in order to increase their absorbing surface. This contrasts strongly with floating leaves, which are usually large, rounded or oval in shape, with their stalks attached near the centre of the lower surface, so that the drag of the stalk does not tilt the leaf and pull it below the water. The leaves of the white and yellow water lilies are instances of this.

Many water plants have both submerged and floating leaves, and when they do the two kinds of leaf are of quite different shapes. The submerged leaves are either long and ribbon-like or finely dissected, while the floating leaves are more or less rounded. The water crowfoot (*Ranunculus heterophyllus*) is an example (Fig. 43). In the arrowhead (*Sagittaria sagittifolia*), which grows on the banks of streams, there

189

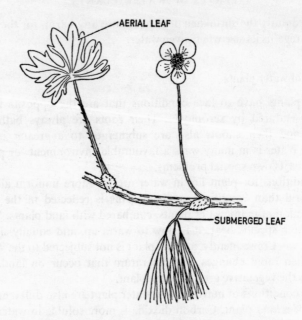

Fig. 43. The water crowfoot (*Ranunculus heterophyllus*)
showing the difference between aerial and
submerged leaves

may be three kinds of leaf on the same plant, particularly if it is
growing by still, shallow water. The submerged leaves are long and
strap-shaped, the floating leaves are oval with notched bases, and the
aerial leaves raised up above the water are the typical arrow-shaped
leaves from which this plant gets its name.

The submerged leaves of water plants bear no stomata, since the
carbon dioxide is absorbed directly in solution, in the form of
bicarbonate ions. Floating leaves have no stomata on their lower
surfaces, but normal stomata are distributed over their upper surfaces.
The upper surface may have a thick, waxy cuticle, to make wetting
difficult, and in some cases the floating leaf is turned up at the edges,
as may be seen in the great leaves of the giant water lily, *Victoria
amazonica*, which grows in the Amazon. This magnificent plant
may have leaves that measure two yards across. Its pollination by
cockchafers has been discussed earlier (Chapter 7, page 142).

The stems of water plants have to bear little weight, for part or all
of the plant is supported by the water. It is therefore not surprising to

190

find that the development of wood is small; some of the most successfully adapted water plants, such as the completely submerged hornwort (*Ceratophyllum demersum*), have no xylem at all, the wood being represented by a hole where the cells that normally give rise to xylem have disappeared. The phloem, on the other hand, is well developed. This, again, is not surprising. Since the plant can absorb water all over its surface, there is no need of water-conducting tissue, but organic compounds produced in the leaves must, as usual, be carried to all parts of the plant.

If wood is developed in a water plant, it usually occupies a position near the centre of the stem, instead of being dispersed near the periphery as in most plants. This is sensible, for a water plant has to resist a pulling strain, due to the current in the water, rather than a bending strain. The disposal of the woody tissue, therefore, is as in a root rather than a stem.

Another feature of water plants is the abundance of air spaces in the cortex of the stem and root. The gas in these spaces consists largely of oxygen, which is given off during photosynthesis. The spaces are continuous down the stem, and into the root, but there may be a series of diaphragms formed across the nodes, which are perforated by small holes which will allow gas to pass but not water. This is a precaution against part of the stem being broken off, either accidentally or in asexual reproduction by fragmentation. Should this happen, the internal spaces do not become flooded.

One of the disadvantages of an aquatic habitat is that water cuts down the intensity of the light very quickly; even at a depth of a few inches the reduction is considerable. This means less light for photosynthesis. To make the most of the available light, the chlorophyll-containing cells must be near the surface of the plant. It is no uncommon thing in water plants to find chloroplasts in the epidermis. In land plants, the only epidermal cells that contain chloroplasts are the guard cells of the stomata.

Reproduction of water plants

With the general exuberance of vegetative growth that is found in water plants it is not surprising that vegetative methods of reproduction should be well developed and prolific Some water plants seldom reproduce by seeds. A good instance of this is seen in the Canadian pondweed (*Elodea canadensis*). This small but ubiquitous plant was introduced into this country in 1842, having reached Ireland from Canada six years previously. It spread like the plague, and indeed 'plague' was an apt word for it, for it soon began to block the English canals, which at that time were an important means of transport. It accomplished this feat entirely by vegetative reproduction, for *Elodea* is dioecious (having male and female flowers on separate plants), and the male plant is seldom seen in Britain. *Elodea* reproduces by fragmentation, branches breaking away from the main stem and starting life on their own. This method of reproduction is so effective that the lack of seeds does not appear to inconvenience the plant at all.

The mangroves come once again into the picture. These semi-aquatic plants are evergreen trees that inhabit tidal zones on tropical coasts, particularly in Florida and Far Eastern countries such as Malaya and Sumatra. They belong to a number of quite unrelated families which by virtue of their habitat have evolved many common characteristics such as 'breathing roots' and seeds that germinate while still attached to the parent plant (see the explanation of 'viviparous' on page 172). They are an ecological group, not a taxonomic one, and show a good example of what is known as evolutionary convergence—organisms which, though they differ widely in descent, have come to resemble one another through their adaptations to the same habitat. We see a similar thing, zoologically, in the resemblance between a shark (fish), an ichthyosaurus (reptile), and a whale (mammal).

As we have seen mangroves grow in swampy land over which the tide flows daily, their feeding roots are buried in the estuarine mud, and from the lower parts of their stems arise an often complex system of stilt roots, so that the plant looks as if it is propped up by an irregular scaffolding. These stilt roots entrap sediment carried by the tide, so that the level of the land in a mangrove swamp is always rising. In this way mangroves may be an important factor in countering coastal erosion.

Another prominent feature of mangroves that I have already mentioned is the way air is carried down to the submerged root system by means of 'breathing roots' or pneumatophores. The need for breathing roots arises because the swamp water that bathes the roots of mangroves is deficient in oxygen, so that the roots have difficulty in getting enough to supply their respiratory needs. Waterlogged soil is always low in oxygen, and in a mangrove swamp the lack is intensified by the amount of rotting organic matter that collects on the bottom of the swamp. The bacteria that are always busy on the breakdown of organic matter also need oxygen, and are therefore competing with the roots for supplies of air.

The possession of 'geniculate' or 'knee-like' breathing roots is not a monopoly of the mangroves, for certain other swamp plants have them. A well-known example is the bald cypress (*Taxodium distichum*).

Climbing and twining

The climbing habit is another adaptation that is found in a varied assembly of plant families. There are many examples in the British flora but the habit is seen at its best in the tropics, and especially in

Fig. 44.
A prickle on the
stem of a blackberry plant

the tropical rain forest where climbing plants, or lianes as they are often called, are an important element in the flora.

Plants climb by various means. Darwin, whose book on climbing plants should be read by everyone interested in plant life, recognized four main classes of climbers. These are hook climbers, twining plants, plants with sensitive climbing organs such as tendrils, and root climbers.

Hook climbers are the simplest and least adapted of the four groups. They are really scrambling plants rather than climbers: their relation to the climber is that of the hillside walker to the alpine mountaineer. They possess prickles that curve backwards like those of the bramble, acting as grappling irons and allowing them to cling to other plants while they scramble over them (Fig. 44). Our only British hook climbers are the blackberry (*Rubus fruticosus*), the raspberry (*R. idaeus*), some species of rose, and some of the bedstraws (*Galium*), but in the tropics the habit is widespread. Some tropical hook climbers have stems or leaves modified as hooks, instead of the humble prickles of the British species (Fig. 45).

Twining plants climb by twisting their stems round a support, the runner bean being a familiar example. Twining plants are better adapted as climbers than the hook plants. The twining habit is brought about by a phenomenon called 'nutation'; this occurs in most plants, but is very noticeable in the twining plants. Nutation is a circular motion of the growing tip of the stem as it grows upwards. The tip executes a wide spiral, seeking for a support. When it finds a

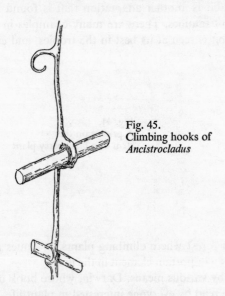

Fig. 45.
Climbing hooks of
Ancistrocladus

vertical stick or other suitable object, the stem winds itself closely round it, so that the plant is supported.

Many twining plants twine always in the same direction. The hop, the black bindweed and the honeysuckle (*Lonicera periclymenum*), for instance, always twine themselves clockwise round their supports, while the bellbine (*Calystegia sepium*), a plant closely related to the black bindweed, and the runner bean, twine anticlockwise. The reasons for these preferences are not known. Some twining plants seem to have no preference as to their direction of twining.

The growth of a twining plant about its support seems to be brought about by the tendency of stems to grow away from gravity.

When the tip of the stem flops over under its own weight the gravitational response causes the lower side to grow faster than the upper, so that the tip turns up again. This constant turning upwards, combined with the strong nutation, produces the spiral growth that results in the stem encircling its support. There is no evidence that any other factor is involved.

Plants with sensitive climbing organs, such as tendrils, are the most specialized of the climbing plants. A tendril is sensitive to touch. When the tendril of a pea plant comes into contact with a possible support, such as the stem of another plant, it bends towards the side which first touched the support. This brings more of the tendril into contact, and so intensifies the response, so that the tendril wraps itself firmly round the object to which it is clinging. This response to a contact stimulus is called 'haptotropism'.

When the tendril has taken a firm hold on its support, further changes occur in its lower part. It becomes spirally coiled like a spring, and this is what is formed; it enables the tendril to give a little as its support is shaken by the wind, so that it is not torn from its hold. Somewhere in the course of the spiral the direction of coiling becomes reversed; this is a natural consequence of trying to coil up something that is fixed at both ends. As the tendril grows older, the coils often become woody.

The bending of a tendril towards a support is brought about by a redistribution of auxin, the substance that is responsible for the reaction of roots to gravity, and of stems to light and gravity (Chapter 4). When contact is made with a solid body, auxin collects on the side away from the support; the tendril therefore grows faster on that side, so that it turns towards the support.

It has been shown that the stimulus that evokes the response in tendrils is not so much touch as friction. If the tendrils of white bryony are gently stroked on one side they curl towards that side, even when the stroking is stopped. If they are merely touched, without the stroking movement, no response occurs. A tendril that has been carefully fixed in contact with a support, so that no movement of the tendril is possible, shows no bending. It is the *kind* of stimulus that matters, not the time through which it acts.

Tendrils are insensitive to contact with liquids. Drops of water dripping on to them do not produce any curvature, and even mercury, the heaviest of liquids, more than thirteen times as dense as water, has no effect. This lack of sensitivity to liquids is an advantage to the plant, which has to stay out of doors in all weathers. The tendrils are able to keep to the job in hand without being side-tracked by every shower of rain.

The part of the plant that is modified to form tendrils varies in different species of plants, but most commonly it is the leaf, or part of the leaf, that is adapted for climbing. This can be seen particularly well in many of the Leguminosae, an order that has produced many tendril climbers. The leaves of leguminous plants are commonly pinnate—that is, divided into a number of leaflets on either side of a common stalk. In the vetches (*Vicia*) one or more of the end leaflets have lost their blades and have turned into tendrils. In the sweet pea (*Lathyrus odoratus*) all the leaflets but one pair have gone, the remainder of the leaf having changed into a tendril. In order to compensate for the loss of the photosynthetic surface of the leaf, the stipules (two leaf-like appendages formed where the leaf stalk meets the stem) have enlarged enormously and taken over the functions of the leaf. The stem of the sweet pea also has a pair of flattened wings running down each side. These are green, and assist with the work of photosynthesis.

The traveller's joy or old man's beard, together with other species of *Clematis*, have adopted a different method of tendril formation. Here the blade of the leaf remains unaltered, but the plant climbs by means of its sensitive leaf stalks. The garden 'nasturtiums' (*Tropaeolum*) also climb in this way.

In *Gloriosa* and *Littonia*, tropical members of the lily family, the tips of the leaves act as tendrils (Fig. 46).

Sometimes the tendril may be a modified branch, as in the grape vine (*Vitis vinifera*), and members of the Passifloraceae, such as the passion flower (*Passiflora edulis*) and the sweet calabash (*P. maliformis*). In the white bryony the exact nature of the tendrils is uncertain. They arise, accompanied by short branches, in the angles between foliage leaves and the stem, and some botanists consider that they are the modified first leaves on the branches, while others look on them as extra branches.

Some tropical lianes climb by means of recurved hooks at the ends of modified inflorescences. These project from the stem and when they catch a support clasp it firmly and become lignified. This specialized habit is seen in a number of different families, including the lily family (*Hugonia*), the periwinkle family (*Landolphia*), and the bedstraw family (*Ourouparia*).

Lianes which climb by means of adventitious roots include the ivy (*Hedera helix*). The stem produces bunches of roots at intervals; these are known as adventitious roots because they are not part of the primary root system (Fig. 13). These climbing roots are peculiar in that they grow away from the light (most roots are insensitive to light), so that they readily find their way into crevices in the support, whether it be the rough bark of a tree or an old wall. The climbing roots are also unusual in not being sensitive to gravity, whereas most roots have a strong tendency to grow downwards. This again enhances their utility as organs of attachment, since they grow out horizontally and so make contact with the support.

Climbing roots are found in various tropical lianes, including many

Fig. 46.
Leaf tendril of
Gloriosa

of the Araceae, and in many climbing ferns. In some species of *Philodendron* (Araceae) that grow perched on the branches of trees, the aerial roots twine about the branches of the supporting tree as they descend towards the ground.

The essence of the climbing habit is that it enables a plant to gain considerable height without having to expend time and energy on strengthening its stem. This is seen to perfection in the lianes that occur in such profusion in a tropical rain forest. These great vines, their stems as thick as a man's thigh, may ascend to the top of the tallest tree, come down to earth again, and climb another tree, attaining a fantastic length as they do so. The stem of a climbing palm measured by Treub in 1883 was more than two hundred and forty yards long—over a furlong! This was perhaps exceptional, but a tropical liane seventy yards long is not exceptional.

Since a liane is supported by the tree against which it grows, it has little need of strengthening tissue to keep it rigid. In lianes, fibres are developed poorly or not at all. This is an advantage, because a fibre takes a long time to grow; by putting its energy to better uses, the plant can grow faster. Water conduction, on the other hand, is liable to be a problem, for the liane has to carry water a long way through a relatively thin stem. So the xylem is well developed, with particularly wide vessels.

Another essential for the stem of a liane is flexibility, for it must be able to twist and turn as it makes its way to the top of a tree. The traditional arrangement of the xylem in the stem of a woody plant, which is in the form of an unbroken ring, is unsuitable. This is taken care of in the secondary growth of a liane, which shows varying anomalies all designed to combine flexibility with a good development of xylem for conduction. *Clematis*, for instance, instead of forming secondary xylem and phloem all round its circumference, forms thin-walled tissue at intervals, so that the xylem remains cut into separate bundles throughout the life of the plant. The vascular structure of some tropical lianes is exceedingly complex, as is shown in many of the Leguminosae such as *Aristolochia* and *Bauhinia*.

When a plant is grown in poor light it shows a number of symptoms that collectively make up the condition known as 'etiolation'. Among these symptoms are a long weak stem and long internodes (an internode is the length of stem between two consecutive leaves). This is exactly what we have in most climbing plants. It has been suggested that the climbing habit might have arisen out of etiolation, the efforts of the plant to rise into the light and air producing a weakened stem which had to have some support. Bearing this in mind, it might be no coincidence that the liane is especially a plant of the tropical rain forest, where the thick canopy of the trees would make for etiolation in a plant growing up from the ground. If the liane has indeed arisen from an etiolated seedling, it would not be the first time that plants have found that the uses of adversity are sweet.

The lianes lead us naturally to the next line in adaptation to environment, shown by the epiphytes. These are plants that have forsaken the earth altogether, spending their lives perched on the limb of a tree, or in some other inaccessible place; we have seen that some epiphytic orchids even grow on telephone wires.

Adaptations by epiphytes

Some plants may find themselves epiphytes by chance, their seed happening to fall in the crotch of a tree and germinate there. I have seen the white dead-nettle growing, apparently quite comfortably, on a branch of an oak. Presumably some windblown soil had collected in a hollow and was sufficient to meet the humble needs of the plant. The tops of pollarded willows often bear small flowering plants growing epiphytically. Such strays as these, however, are only casual epiphytes. A true epiphyte—a plant that normally grows mounted upon another—is morphologically and physiologically adapted for its peculiar mode of life. If it were not, it could not survive.

Epiphytes do not grow well in cool climates. In Britain we have only one true epiphyte amongst the higher plants; this is the polypody (*Polypodium vulgare*), a fern that grows on trees as well as among rocks. All our other epiphytes are lower plants; mosses, algae, and lichens. Epiphytic flowering plants flourish where the climate is kinder, and are to be found in abundance in the tropical rain forest, and even more in the montane and subtropical rain forests. The life of an epiphyte is precarious even under the best conditions, and to expect them to cope with a climate such as ours in addition to their other difficulties would be asking a little too much.

To become established in the dizzy height of a tree in a rain forest, perhaps two hundred feet or more above the ground, an epiphyte needs an efficient system of seed dispersal. Only two agencies can accomplish it regularly and without fail. These are wind and birds, and it is therefore only to be expected that nearly all epiphytes rely on one or the other of these two methods of seed dispersal.

Fleshy fruits are common among epiphytes, and many contain a sticky pulp that makes a bird wipe its beak on the bark of trees, so rubbing off the seeds. This mechanism has been described for our own mistletoe, which is a partial parasite rather than an epiphyte though it grows on trees. It is found also in the tropical *Rhipsalis*, and many of the epiphytic Bromeliaceae. The seeds of the Bromeliaceae are often deposited by birds on telephone wires, where plants may become established.

The spores of ferns are extremely small and light, and are easily carried to considerable heights by the wind. The same may be said of the seeds of orchids, which may be so small as to form a powder. *Stanhopea oculata* has seeds each of which weighs only one ten-millionth of an ounce. Seeds with parachute mechanisms are common in tropical and subtropical epiphytes.

The Spanish 'moss' (*Tillandsia usneoides*) is a remarkable flowering plant found in tropical and subtropical America. In habit it resembles the lichen *Usnea*—hence its name. It hangs in long grey festoons, the base of the stem being wound round the limb of a tree. As the stem grows downwards the older parts die off, leaving the central strand of fibrous tissue intact, looking rather like horsehair. The plant is covered all over with scaly hairs for absorbing water that trickles over it from above (Fig. 47). Flowers are seldom formed, and the plant

Fig. 47.
Tillandsia usneoides,
an epiphyte from
South America

depends for its dispersal partly upon the wind, which tears off parts of the beard-like mass, and partly on birds, which carry it off to use as nesting material.

When the seed of an epiphyte has been carried to its perch and germinated, the next problem is how to stick there. This is nearly always accomplished by means of the root system. *Aeschyanthus*, an orchid, has a seedling furnished with a flattened disc and well provided with root hairs by means of which it gains a hold and hangs on until established. In many members of the Araceae, Orchidaceae and other families there are two kinds of roots, some being nutritive while others serve purely for attachment. In the Bromeliaceae, which are highly specialized epiphytes, water absorption is carried out by hairs on the leaves rather than the roots, which are entirely specialized for attachment. They are stiff and wire-like, and absorb little or no water.

The seed having been carried to a suitable place on the trunk or branch of a tree, and the plant having in due course managed to establish itself, the real troubles of the epiphyte begin. Entirely divorced from the soil as it is, it has somehow to collect enough water to maintain life, and with the water sufficient mineral salts to satisfy its needs. This is by no means easy, and it is particularly in connection with their water relations that we find the most vital adaptations to epiphytic life.

The botanist Schimper, writing towards the end of the nineteenth century, recognized four classes of epiphytes, each showing a different grade of epiphytic specialization. These were: 'proto-epiphytes', 'nest and bracket epiphytes', 'tank epiphytes', and 'hemi-epiphytes'. In the proto-epiphytes, there are no specialized structures for the collection of water and soil. Many show xeromorphic characters such as succulent leaves, and water-storing organs are common. Many orchids belonging to this group have aerial roots with a special water-absorbing tissue, the velamen. Its functions have been outlined on page 150 near the beginning of Chapter 8. This velamen is on the outside of the root, and consists of large empty cells often with spiral or reticulate thickening bands on their walls, which are provided with holes leading from one cell to another. The velamen acts in a similar way to blotting paper in absorbing water, which readily enters the empty cells. Beneath the velamen, the exodermis of the root is modified in a similar way to the endodermis, the cells having thickened wall, except for a few which remain thin-walled, like passage cells (Chapter 2). One can hardly doubt that this is connected with the absorption of water by the velamen.

The proto-epiphytes are the least specialized group of epiphytes, and are liable to suffer more than any others from the effects of intermittent drought.

In the nest and bracket epiphytes, the structure of the plant is adapted in such a way that it can accumulate debris, thus gaining a private supply of humus. In nest epiphytes, the tangled roots resemble a bird's nest. Humus and rotting debris collect among the roots, which serve as protective structures under which ants build their nests, thereby adding to the supply of humus.

In bracket epiphytes, brackets are formed by some or all of the leaves, which lie against the trunk of the tree. Humus collects in these brackets, the accumulation often being assisted by nesting ants.

The remarkable stag horn ferns (*Platycerium*), which grow in the forests of Malaya, are good examples of bracket epiphytes. They possess two kinds of leaves. One kind stands erect, and is known as the 'mantle'. In *P. grande*, the lower part of the mantle clings to the trunk of the tree on which the fern grows, while the upper part spreads out and forms a bracket in which humus collects. In *P. bifurcatum*, the whole of the mantle is closely pressed to the supporting trunk; the only humus that collects is from the decay of the old mantle leaves, with the possible addition of bark from the tree. In both species the other kind of leaf, sometimes called the 'sporangiophore,' hangs downwards. It is much branched, like the antlers of a stag (hence the name of the plant), and bears sporangia towards its tips.

There is a strong connection between ants and epiphytes, amounting in some cases almost to symbiosis;* 'commensalism' is the term generally used for such associations, where the two partners live together but are also free to go their own ways at times. In *Myrmecodia*, an epiphyte from Malaya, there are large stem tubers which contain cavities, connected by galleries, which are inhabited by ants (Fig. 48). The main function of the tubers seems to be water storage, and it is difficult to understand why the cavities, with their connecting galleries, should be formed at all, for the ants do not appear to do anything for the plant.

Hydnophytum is another epiphyte found in the forests of New Guinea and Fiji, that has ant-inhabited stem tubers similar to those of *Myrmecodia*. Both *Hydnophytum* and *Myrmecodia* belong to the bedstraw family (Rubiaceae).

Tank epiphytes

The tank epiphytes are found only in the Bromeliaceae, a family that shows a high degree of specialization for the ephiphytic mode of life. The Bromeliaceae are found in tropical America and the West Indies. The remarkable tanks by which the Bromeliaceae catch and hold water are formed by rosettes of fleshy leaves which fit closely at the base, and form a kind of funnel above. Water collects in the

* See Glossary and opening paragraphs of chapter 14.

pitchers so formed, together with debris of various kinds, including dead and living plants and animals. The flora and fauna of bromeliad tanks is extremely varied. Some Brazilian species of bladderwort (*Utricularia*) are found only in the tanks of the Bromeliaceae. Among the fauna of the tanks are large numbers of mosquito larvae and pupae together with other aquatic insects, and frogs regularly breed there.

The bromeliads absorb water from their tanks by means of scaly hairs on their leaves. The leaves themselves contain much water-storage tissue. Along with the water the plant absorbs nutritive matter derived from the humus that collects in the tanks. The roots of tank

Fig. 48.
Stem tuber of *Myrmecodia*.
The tubers are inhabited
by ants

bromeliads are relieved of their absorbing function, and serve only as a means of attachment.

Adaptations by Hemi-epiphytes

The hemi-epiphytes are plants which begin life as epiphytes, but develop long aerial roots which make their way downwards until they reach the soil. Once contact with the soil is made, the problem of finding water is solved. *Coussapoa fagifolia*, a member of the mulberry family (Moraceae) from tropical South America, is quite a large shrub that manages to live as a hemi-epiphyte, while a number of Araceae are herbaceous members of the group.

Examining the epiphytic flora in a tropical forest, one is struck by the fact that the higher up the trees the epiphytes grow, the more pronounced are their xeromorphic characters. Epiphytes on the smaller trees that form the lowest of the three levels usually to be found in a tropical rain forest show little, if any, adaptation to water economy. Living as they do in a constantly humid atmosphere and protected from the wind, water supply offers few problems. On the other hand, epiphytes growing on the topmost branches of the tallest trees need advanced xeromorphic adaptation if they are to survive.

It is interesting to speculate on how the epiphytic habit arose, though nothing certain is at present known about its origins. Some epiphytes probably began as lianes, and ended up by permanently losing their connexion with the soil. This can be seen in the Malayan orchid, *Dipodium pictum*, which starts its life rooted in the soil, but eventually lives as an epiphyte. Some of the Araceae also behave in the same way. It is tempting to note in this connexion that the Araceae, a family noted for its lianes, also contains true epiphytes such as *Philodendron*, which have no connexion with the soil at any time in their lives. Plausible as this explanation may be of the origin of some epiphytes, it certainly cannot account for the origin of the habit in general.

Schimper put forward the theory that epiphytes were originally terrestrial plants growing in wet, shady forests. Yet most epiphytes show great intolerance of shade, so that Schimper's theory can hardly be true except perhaps for a few 'shade epiphytes' that grow at low levels in the forest. A more probable theory is that the epiphytic Bromeliaceae, at any rate, arose from ancestors that grew on the edges of South American deserts. The same theory would fit certain other epiphytes, such as the epiphytic Cactaceae. An attractive part of this idea is that the plants evolved many of the characters useful to them as epiphytes while still rooted in the ground.

Epiphytes are in no sense parasitic on the trees that bear them; they use their 'hosts' purely as perches on which to rest. In spite of this, however, there is a strong correlation between certain tree species and the epiphytes that inhabit them. In the Philippines, for instance, the fern *Stenochlaena areolaris* is only found growing on the 'screw pine' *Pandanus utilissimus*. It was formerly thought that the specificity of certain epiphytes to certain trees was due to physical causes, such as the degree of roughness of the bark. In many cases this is probably so, but recent work has shown that the association is closer than can be explained in terms of physical conditions, and must presumably be due to chemical phenomena of some kind. The epiphyte depends for its water at least partly on the drainage from the surface of the tree. Such water might well contain excretions given out by the host. For example, it is possible to show that the water draining from the branches of the asiatic chestnut (*Castanopsis argentea*) contains tannin. The chemical composition of the outer layers of the bark might also influence epiphytes growing on a particular species of tree.

Adaptations by halophytes

The chemical composition of the environment can have a drastic effect on plant life, and call for extremes in adaptation. No plants show this better than the 'halophytes'—plants that live in salt marshes, around sea coasts, and in salt steppes. In such places there is an excess of salt in the soil, which makes the absorption of water by roots extremely difficult.

We saw in Chapter 3 that water enters the roots of a plant because the osmotic pressure in the roots is higher than that of the soil water —or, more accurately, because the diffusion pressure of the soil water is higher than it is in the roots. If the osmotic pressure is high, the diffusion pressure of water is correspondingly low. Water containing a great deal of salt in solution has a high osmotic pressure; the diffusion pressure of the water will be decidedly low, so that its entry into the roots of plants will be impeded.

Halophytes can counter this effect, at least to some extent, by raising the osmotic pressure of their cell sap. Measurement of the osmotic pressure in the roots of certain halophytes has confirmed that this occurs. The osmotic pressure in the cells of normal plants is usually from five to ten atmospheres, and seldom rises to twenty atmospheres except in storage organs containing a great deal of sugar. Measurements of the osmotic pressure in the cells of *Atriplex confertifolia*, a plant growing near the Great Salt Lake in America, gave the astounding figure of one hundred and fifty-three atmospheres.

Osmotic pressures of this magnitude may give rise to physiological problems that have to be solved if the plant is to live successfully. Photosynthesis, for instance, may be affected if the osmotic pressure in the leaves reaches a high level.

Since halophytes find it difficult to absorb enough water, it is natural that they should seek to conserve what they have. That is why we find that halophytes usually possess xeromorphic characters of one kind or another. Although they may not be xerophytes—saltmarsh plants certainly are not—they are growing under conditions of physiological drought which may be just as exacting as actual drought. It is therefore quite usual to find halophytes with thick cuticles, sunken stomata, fleshy leaves and all the other trappings of xeromorphy.

Calcicoles and calcifuges

Plants growing on alkali plains such as are found in some parts of America suffer the same disadvantages as halophytes from soil water with a high osmotic pressure, and may show the same kinds of adaptations. These plants have another problem to face in the alkalinity of the soil. The same problem, in a lesser degree, confronts plants growing in chalk. Most fertile soils are slightly acid, but in places where the chalk rises to the surface, as in the North and South Downs of England, it may be distinctly alkaline. Where this happens, a distinctive flora occurs, known as the chalk flora. Typical chalk plants are the purging flax (*Linum catharticum*), the yellow-wort (*Blackstonia perfoliata*), the centaury (*Centaurium minus*), and many of our British orchids.

Some plants will only grow on a calcareous soil, and are known as 'calcicoles'. Others shun the chalk, growing only on acid soils such as a sandy heath or a moor covered with peat; they are called 'calcifuges'. Between these two extreme types are the plants which seem to be indifferent about the reaction of the soil, growing equally well on either type. This fact has been known for a long time, but it is not yet certain what it is that makes calcicoles love the chalk, or calcifuges hate it. It is probable that some calcicoles grow on chalk because they like the physical conditions prevailing on a limestone soil. But this explanation will not work for all calcicoles.

An examination of some plots on a Lower Greensand soil that had been heavily treated with lime some time in the remote past produced some interesting results when the plants growing there were compared with the normal Lower Greensand flora. The limed plots did not differ in the slightest respect from the neighbouring soil except in their lime content, which was as high as 4·5 per cent, compared with a barely detectable trace on the characteristic Lower Greensand soil. Yet several strong calcicoles, such as the salad burnet (*Poterium sanguisorba*) and the hairy hawkbit (*Leontodon hispidus*), grew on the limed plots and nowhere else. It seems difficult to avoid the conclusion that these were chemically determined calcicoles, which needed the actual calcium carbonate, and not merely its effect on the soil.

Why these plants should have such an urgent need for chalk is uncertain. It is true that calcium is an element essential to the lives of all plants, but the small amount of calcium needed could have been obtained from almost any soil. It must be something more subtle than this.

The calcifuges, intolerant of lime in the soil, grow best where conditions are acid. The bog-mosses (*Sphagnum*), the ling (*Calluna vulgaris*), and the purple bell heather (*Erica cinerea*), are examples. Some are so intolerant that the presence of lime in their water actually poisons them, so that there can be no doubt that their calcifuge habit is chemically determined. Again, the reason for this intolerance of lime is not known with certainty.

Photoperiodism

The number of hours of daylight between dawn and dusk have an important influence on the flowering of certain plants. This was first observed by a Frenchman called Tournois, who noticed that the hemp, a plant that normally does not flower when grown in northern latitudes, where there is a long daily period of illumination, could be made to flower if the length of the day was artificially cut down. At about the same time, Klebs found that a species of houseleek, *Sempervivum funkii*, could be made to flower during the winter if it was given extra illumination by means of electric lamps. The hemp plant required a short day if it was to flower, while the houseleek required a long one.

Soon, other plants were found to be equally sensitive to the length of the daily period of illumination. In 1920 Garner and Allard, who were working on new varieties of tobacco, found that a variety called 'Maryland Mammoth' would not flower in the open at Washington D.C., though it flowered during the autumn and winter if it was grown in a greenhouse. This was unfortunate, for Garner and Allard wished to cross 'Maryland Mammoth' with other varieties, and could not do so because of its lack of flowers. When they tried shortening the day length for 'Maryland Mammoth' by putting the plants in the dark after a certain number of hours each day, they were rewarded with an abundance of flowers, and were able to make their cross.

The success of this experiment led Garner and Allard to try the technique on other plants that normally did not flower in their locality, and they found that the response to a shortened day was by no means confined to 'Maryland Mammoth' tobacco. In this way they uncovered what appeared to be a general phenomenon in plants, to which the name 'photoperiodism' has been given.

Plants can be divided broadly into three groups according to their photoperiodic responses. Short-day plants produce most flowers if they are given daily light periods of twelve hours or less; such plants include hemp, 'Maryland Mammoth' tobacco, many tropical flowering plants, and many of the plants of temperate regions that normally flower in spring or autumn. Long-day plants, on the other hand, flower best if given daily light periods of twelve hours or more. Long-day plants include *Sempervivum funkii*, the henbane (*Hyoscyamus niger*), the potato, wheat, and all the plants of temperate latitudes that normally flower in the summer. Besides the short-day plants and long-day plants, there is another class known as 'day-neutral', in which the flowering does not seem to be governed by the length of day. This group includes the tomato and the dandelion.

It should be noted that this physiological classification according to the photoperiodic response cuts across the normal botanical classification into families and genera. The photoperiodic responses of two closely related plants may be quite different. The tobaccos, for instance, are all varieties of one species, *Nicotiana tobaccum*, yet the variety 'Maryland Mammoth' behaves differently from most other varieties. The tomato and the potato both belong to the same family (Solanaceae), yet the potato is a long-day plant while the tomato is day-neutral. There seems to be no connection at all between degree of botanical relationship and the photoperiodic response.

The above classification is only a rough one. Each individual plant has its own characteristic photoperiod, which is exact; the critical photoperiod for 'Maryland Mammoth' tobacco, for instance, lies between thirteen and fourteen hours, and a daily period of illumination longer than the critical photoperiod completely inhibits flowering. The henbane, a long-day plant, has a critical photoperiod of from ten to eleven hours, and will not flower if given a daily period of illumination *shorter* than this. Thus, 'Maryland Mammoth' tobacco and the henbane overlap; a photoperiod of twelve hours will allow both to flower.

For short-day plants, it seems that it is a long period of darkness, rather than a short day, that is the deciding factor in determining flowering. Long-day plants, on the other hand, require no period of darkness; they will flower even if kept under continuous illumination.

A plant does not have to be kept under the right conditions of illumination all its life in order to flower, for in many instances a surprisingly brief exposure to the right photoperiod is all that is needed. The record is probably held by *Xanthium*, a short-day plant, which will flower in spite of being kept at the wrong photoperiod, provided that it is allowed just one short day, followed by a long night. This stimulation to flowering by giving a short exposure to the right photoperiod is known as 'photoperiodic induction'.

The effect of light in stimulating long-day plants to flower, or in inhibiting flowering in short-day plants, is not so much dependent on the intensity of the light as on its duration. Exposure to a 100-watt bulb at a distance of eighteen inches is quite sufficient for most plants, and some are far more sensitive. *Xanthium*, for instance, can be prevented from flowering if exposed to 0·3 foot-candles—equivalent to a single 100-watt bulb placed at a distance of five hundred feet!

From this, it is evident that the photoperiodic effect of light cannot be due to photosynthesis. On the other hand, photoperiodic induction is not effective if the air is freed from carbon dioxide during the period of induction.

It is not necessary to illuminate the whole plant in order to secure a photoperiodic effect, and experiments have shown that it is the leaves that must be illuminated, and not the buds. Spinach (*Spinacea oleracea*) is a long-day plant, but it will flower under short-day conditions provided that the leaves are given long days; reversal of this process, with the buds illuminated while the leaves are kept dark, produces no effect.

The effect of illuminating the leaf is shown clearly by grafting experiments, for a leaf that has undergone photoperiodic induction will induce flowering when grafted on to a plant that has been kept under the wrong photoperiod. Even a detached leaf can receive photoperiodic induction, and will produce its effect when grafted.

These experiments on grafting suggest strongly that some substance is produced in the leaves as a result of exposure to the correct photoperiod, and that this substance then travels from the leaves to the growing points, stimulating flower initiation. Further experimental work has made this virtually certain, though in spite of many attempts, no substance has ever been extracted from leaves which is capable of stimulating flowering. None the less, the name 'florigen' has been given to this hypothetical substance.

The geographical effects of photoperiodism are very striking. In the tropics where days are relatively short we have the short-day plants, and those short-day plants that are found in temperate regions flower during spring or autumn. Long-day plants, on the other hand, are the typical summer-flowering plants of temperate latitudes, where the long summer days suit them.

Did short-day plants gain a hold in the tropics because they like short days, or has the short-day habit evolved because these plants are tropical? We cannot answer this. In their adaptation to short or long days, however, we see another instance of sound design in plants.

The calcicole or calcifuge disposition, and photoperiodism, are just two of many physiological adaptations to a particular environment. There are many others. Physiological specialization may not be as obvious or as exciting as the modifications of structure that enabled plants to become lianes or epiphytes, but they are just as important. The world is full of plants; so full that every individual plant has to fight for its existence from the time it starts off as a hopeful seedling to the time it sets fruit and, if it is an annual, dies leaving only its seed. Any morphological or physiological quirk that can enable a plant to steal a march on its fellows must be exploited. That is an excellent thing, for it has given us all the wonderful variety of plants that we see and hear about.

There are very few vacant places for plants in the world today. There may be a few vacancies for arctic-alpine plants on the slopes of Kilimanjaro and other equatorial mountains, if they could get there and develop adaptations that would carry them a little higher than their competitors; but such vacancies are few and scattered. Meanwhile, plants in every situation must hang on to such advantages as they have, striving after every small improvement that may confer even the most tenuous advantage over their competitors. In Nature, to cease from striving means death for the individual and extinction for the species.

12 · Parasites and saprophytes

Most plants obtain all the carbon they require from carbon dioxide in the air, and their mineral needs are supplied by their roots, but some have adopted other means of nourishment. The great group of fungi have eschewed chlorophyll from the start; they procure all their carbon from rotting organic matter, or by sucking the vital juices of a living host. As far as we know, the fungi never had chlorophyll in their ancestry (unless we believe, as some still do, that they originally evolved from the algae, or primitive water plants).

There are a few flowering plants that have adopted the dietary habits of the fungi, living as parasites on other plants, or finding their sustenance in the rotting organic matter in woodland detritus or elsewhere. These differ from the fungi in being descended from ancestors that possessed chlorophyll; indeed, some parasites—to be strictly accurate, partial parasites—still have chlorophyll in their leaves, and can synthesize part or all of the carbon compounds that they need, depending on their hosts only for minerals. The mistletoe (*Viscum album*) is such a one.

The appearance of the mistletoe is familiar to everyone. It is found growing as a partial parasite on the branches of a great variety of deciduous trees, especially the apple. It is also seen occasionally growing on conifers, particularly the pine and larch, and, very rarely, on evergreen broad-leaved trees. A number of semi-religious myths surround the mistletoe. Its association with Christmas derives from ancient pagan times. The Druids set great store by oak mistletoe, possibly because it so seldom attacks the oak that oak mistletoe is rare.

Partially parasitic life of mistletoe

The peculiar seed dispersal mechanism of the mistletoe has already been described (Chapter 9). When the seed has been fixed to a branch

211

of a tree by the glutinous slime in the fruit, it does not germinate at once; it must wait until the increasing temperature and longer days of spring allow germination to begin, for besides warmth the seed of the mistletoe needs plenty of light before it can germinate. When it does so, a small white cylinder resembling a young root emerges from the seed coat. This is not, in fact, the root, for the mistletoe has no true roots; it is the 'hypocotyl', or the part of the axis that lies below the cotyledons or seed leaves.

At first the growing hypocotyl grows upwards against the force of gravity, but it soon loses its power of gravitational response and starts to grow away from the light. This causes it to bend towards the branch on which it is growing, and with this it soon makes contact. The tip then spreads out into a circular, sticky disc which adheres closely to the surface of the branch. From the centre of the lower side

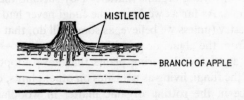

Fig. 49. Section of part of an apple branch,
parasitized by mistletoe

of the disc a peg-shaped sucker emerges. This grows into the cortex of the host, the disc providing a firm base from which the peg can force its way through the host tissues. Growth of the sucker continues until it reaches the wood. Strands of vascular tissue then develop in the sucker, connecting with the wood of the host. This business of primary infection occupies the mistletoe for the first year of its life.

The first pair of leaves appear at the beginning of the second year of growth, which is at first extremely slow; as the plant gets older, however, growth speeds up, and the plant acquires its typical bushy form.

While the mistletoe bush is developing on the outside of the tree, much is also happening unseen within the cortex (Fig. 49). Outgrowths arise from the original sucker and run along the cortex of the host, beneath the bark. These new suckers run, in the main, along the length of the branch; they do not encircle and strangle it. Here and there the suckers produce buds, from which new mistletoe bushes grow, so that from one seed many mistletoe plants are ultimately formed. It is useless to attempt to free the tree of mistletoe by cutting off the bushes as they appear, for the suckers are buried in the cortex, out of reach of the pruning knife.

The nature of the suckers of the mistletoe is uncertain. Some regard them as modified adventitious roots, but they have none of the morphological characters by which roots are distinguished. It is probably best to look on them as special organs of absorption, developed in response to the mode of life of this peculiar plant. Such organs are often called 'haustoria', a term borrowed from the fungi.

The mistletoe is well supplied with green leaves, and so can fulfil its own carbon needs without recourse to its host. It must depend entirely on its host, however, for water and minerals. These it absorbs direct from the xylem of the host by means of the vascular elements in its suckers. The mistletoe is therefore a partial parasite.

One might be tempted to ask whether the mistletoe is, in fact, parasitic at all. Since it has leaves and can photosynthesize, might it not be a case of exchange rather than robbery, the mistletoe paying in photosynthesized sugars for what it takes in water and minerals? That this is not so was shown by a well-known experiment performed by Molisch in 1920. Molisch cut off all the leaves from a young apple tree that had several mistletoe bushes growing on its branches. In a short time the tree died of starvation, and the mistletoe perished with it. Clearly, the mistletoe was not supplying the tree with carbohydrates. Like all parasites, it took all and gave nothing. This is the distinctive mark of a parasite.

The Loranthaceae, the family to which the mistletoe belongs, contains many semi-parasitic shrubs. *Viscum album* is the only British species, but more than five hundred others are found in temperate and tropical regions. Nearly all live as partial parasites, but a few root in the soil. *Nuytsia*, from Western Australia, grows into a small tree about thirty feet high. *Arceuthobium minutissimum* is an interesting species that is a total parasite growing in the Himalayas. Its host is the Macedonian pine (*Pinus peuke*). The parasite grows entirely inside the tissues of the host, only the flowers appearing above the surface of the bark.

The figwort family (Scrophulariaceae) contains several common partial parasites, including the yellow rattles (*Rhinanthus*), the cowwheats (*Melampyrum*), the red rattle (*Pedicularis palustris*), the lousewort (*P. sylvatica*), and the eyebrights (*Euphrasia*). These are all partial parasites on the roots of other plants.

Other partial parasites

The eyebrights and the yellow rattles are not obligate parasites, for they are capable of growing without a host. The cow-wheats, on the other hand, need to be attached to the roots of other plants if they are to develop properly. All these plants possess chlorophyll, and are able to supply themselves with carbohydrates quite adequately. They grow in pastures and meadows, where there are plenty of grass roots to which they can become attached. The seed germinates normally. Early in the life of the seedling, it produces lateral roots which swell, forming tubercles. If one of these comes into contact with a grass root, it sends out a process which enters the root of the grass, developing vascular tissue through which it can rob the root of the host. It seems probable that the main requirement of the parasites from the host is water, for their poorly developed root systems are inadequate to cope with their transpiration.

The sandalwood family (Santalaceae) consists largely of partial parasites, some of which live on branches of trees, like the mistletoe, while others are root parasites like the eyebright. Our one British representative of the family, the bastard toadflax (*Thesium humifusum*), which lives in chalk and limestone grassland, is a root parasite.

Fully parasitic plants

Tozzia is a curious genus of the figwort family which is fully parasitic when young, becoming a partial parasite later. There are two species, one living in the Alps while the other grows in the Carpathians. In both species the seeds germinate to form an underground rhizome which is completely parasitic on the roots of its host. Later, a flowering shoot is formed with pale green leaves containing little chlorophyll. Having produced its aerial shoot, *Tozzia* lives as a partial parasite until the appearance of its blooms, which resemble those of a large eyebright. Having set seed, the parasite dies.

The broomrape family (Orobanchaceae), which is closely related to the figwort family, consists entirely of parasites. The broomrape family reaches its greatest development in the warm temperate regions of the northern hemisphere, but we have several British species, including the toothwort (*Lastraea squamaria*) and a number of broomrapes (*Orobanche*).

The toothwort is parasitic on the roots of woody plants, particularly

214

the hazel (*Corylus avellana*) and the elms (*Ulmus*). It is found in moist hedgerows and woods, on good soil, and is particularly common on limestone. It produces a thick flowering shoot bearing helmet-shaped white or purplish flowers. The aerial shoot also bears scale leaves which are devoid of chlorophyll.

The underground portion of the plant consists of a thick branched rhizome bearing fleshy scale leaves. The scale leaves are bent back on themselves, the upper part being fused with the lower part to form hollow bags. The inner surface of each bag is covered with glands, the function of which appears to be the secretion of water. These strange leaves appear to be food-storage organs, but other functions have been ascribed to them, most of them highly improbable. It has been suggested that the secretion of water by the glands is a substitute for transpiration. This may well be true, for the toothwort is connected with the main water supply line of its large host, and has only scale leaves through which it can transpire. It has also been suggested that the glands may have an excretory function, getting rid of surplus ammonia, phosphates and sulphates, and that small soil organisms which get into the bags are digested and absorbed by the toothwort. There is no evidence whatsoever for either of these suggestions, nor have they the merit of probability.

The seed of the toothwort is said to germinate only in contact with the roots of a suitable host; presumably it must first be washed down into the soil by rain. On the root of the seedling suckers are formed and these, on touching a host root, grow round it and fix themselves by a number of fine outgrowths. When the sucker has become firmly fixed in position, a process grows out from it into the wood of the host. From then on, the toothwort obtains all its nourishment from the roots of the host plant.

Besides the toothwort, another species of *Lastraea* has become naturalized in a few localities in this country. *L. clandestina* is parasitic on the roots of willow and poplar in damp, shady places. Its native homes are Spain, Italy, France, and Belgium.

There are eleven British species of broomrape, all of which are root parasites. The seeds, which are minute, are washed into the soil by rain and germinate when they come into contact with a host root. On germination, a sucker grows out from the root of the seedling and penetrates the root of the host. A feature of the root parasitism of the broomrapes is the intimate contact between the tissues of the parasite and those of the host. As the parasite grows in the host tissues, the wood of the parasite matches exactly with the wood of the host, the cortex matches the cortex, and even the epidermis of the broomrape merges with that of the host.

From the underground portion of the broomrape arises the flowering shoot. This, like that of the toothwort, consists of a stem bearing scale leaves, without chlorophyll, topped with an inflorescence of flowers.

A parasite of quite a different kind is seen in the dodders (*Cuscuta*), of which there are two British species. A third is parasitic on flax, and is occasionally introduced with flax seed. The dodders belong to the convolvulus family (Convolvulaceae), and are climbing plants, twining round the stems of their hosts and drawing nourishment from them by means of suckers.

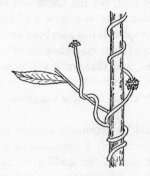

Fig. 50.
The dodder (*Cuscuta europaea*)
growing on willow

The large dodder (*Cuscuta europaea*) parasitizes the stinging nettle (*Urtica dioica*), while the common dodder (*Cuscuta epithymum*) attacks a number of different plants, including the gorses, ling, and clovers (*Trifolium*). The seed of the common dodder is small and very simple in construction. It germinates to form a tiny seedling which is thread-like and with no clear distinction into radicle, plumule and cotyledons. The seedling is capable of a brief period of independent existence, during which the stem shows the strong nutation that is characteristic of twining plants. If the stem meets the stem of a host the remainder of the seedling withers and dies, while the stem twines itself closely round its victim; at this stage, therefore, the dodder has no connection with the soil, and must gain all its sustenance from its host (Fig. 50).

The suckers of the dodder arise from the stem. They first flatten themselves against the host stem and then a peg-like outgrowth arises from the centre of the flattened portion and grows into the host, finally reaching the vascular tissue. Arrived there, it expands a great deal. Although at first the outgrowth consists of undifferentiated tissue, when it has reached the vascular system of the host it develops xylem and phloem which connect up with similar tissues in the host. A connecting link between the vascular tissue of host and parasite is thus set up, and the nourishment of the dodder is assured.

The twining stem of the mature dodder plant is yellow or red, and appears to be without leaves; but close examination shows that there are minute scale leaves here and there. These are without chlorophyll, and incapable of photosynthesis.

Cassytha, a tropical genus belonging to the laurel family (*Lauraceae*), has a similar parasitic habit. There are about fifteen species, all in tropical regions of the Old World.

There is a strong tendency for the parasitic habit in flowering plants to be accompanied by a general simplification in structure. Freed from the necessity of feeding itself, the plant can concentrate its energies on reproduction and particularly on the production of a copious supply of seed, made necessary because a suitable host may not be easy to come by. We see the same thing in animals that are internal parasites. Independent movement is no longer needed, so the legs disappear; often the digestive organs disappear likewise, since the animal is fed by the already digested food of the host. The reproductive organs, on the other hand, are usually well developed, to allow for the risk of failing to find a host which is inherent in the parasitic habit.

The ultimate adaptation to parasitism in plants is seen in the tropical *Rafflesia arnoldi*, which is found in Malaya, parasitizing the roots of vines. In *Rafflesia*, the vegetative body is reduced to a series of filaments, like the hyphae of a fungus, which ramify in the tissues of the host plant. Only the flowers are normal. They are produced directly from the roots of the host, so that they appear to spring from the soil. *R. arnoldi* has the largest flowers in the plant kingdom, as described on pages 165–166. They are of a dull, reddish colour, and have a smell resembling that of putrid meat—hence the name 'stinking lily' by which the plant is sometimes known. The flowers, as their smell might suggest, are pollinated by flies. The peculiar method of fruit dispersal, by the tread of elephants, has already been mentioned.

Saprophytes

Saprophytes are plants that draw their nourishment from decomposing organic matter. Saprophytism is a common habit among the fungi, but in flowering plants it is rare. Why this should be so is uncertain. At first sight, there is much to be said for the saprophytic habit. Organic matter is plentiful in the detritus that carpets the floor of most woods, and a saprophytic plant has no problems about sunlight, or about root competition from other plants in getting its food.

The most probable reason why few flowering plants have adopted saprophytism is that they lack the enzymes needed to break down organic matter into a state in which it is readily taken in by the roots. Although there is evidence that the roots of most plants are capable of absorbing organic molecules to a certain extent, they are unable to do so readily enough to meet their needs in natural conditions, though they may, in certain circumstances, do so in the laboratory (peas have even produced flower buds when grown in the dark on a culture medium containing sugar). It seems that it is easier for plants to become parasites than saprophytes.

It is a significant fact that nearly all saprophytic flowering plants are mycotrophic: they have a fungal partner in their roots (see Chapter 14). The one exception to this is said to be *Wullschlaegelia aphylla*, a saprophytic orchid that is reputed to be fungus-free, though this is uncertain. It is not improbable that the saprophytes rely on the enzymes produced by their fungi to break down the organic matter, or at least relegate to the fungi the task of absorbing it. If that is so, it is at least arguable that the saprophytic flowering plant is in reality parasitic on the fungus.

The best-known British saprophytic flowering plant is the bird's nest orchid (*Neottia nidus-avis*), which grows in beechwoods on chalky soils. This plant gets its name from the dense tangle formed by its roots, resembling a bird's nest. The aerial shoot bears brown scale leaves which contain little if any chlorophyll.

The yellow bird's nest (*Monotropa hypopithys*) is another British saprophyte, found in woods of beech and pine. The whole plant is either yellowish or ivory white in colour, with a waxy appearance, and contains little or no chlorophyll. Its relative, *M. uniflora*, is the 'Indian pipe', found in American woodlands.

13 · Plants of prey

We have already seen (Chapter 7) how important insects are to plants as the principal agents in cross-pollination. Insects and plants have, so to speak, grown up together and the adaptations shown by some flowers to their insect visitors are little less than miraculous. But insects have a more baleful side to their nature. The flea beetles, the aphides and the voracious locusts are among the worst enemies of plants, and the damage they do to plant life during the course of every year is incalculable.

This being so, it is hardly surprising that plants should occasionally reverse the normal predator-prey relationship, and instead of passively serving as food for insects, should strike back and feed on insects themselves. The insectivorous habit has evolved more than once in the plant kingdom, and some remarkable adaptations have been devised by plants bent on making insects their prey.

The sundew (*Drosera rotundifolia*) is a small plant found in bogs and wet peaty places all over Britain. It has a rosette of long-stalked red leaves with circular blades about a centimetre in diameter. At flowering time, which is from June to August, a flower stalk rises to a height of from four to twelve inches from the centre of the rosette of leaves, bearing a cluster of white flowers: altogether a thoroughly insignificant little plant, readily overlooked by anybody who is not searching for it.

Sundew as insect trap

The leaves of the sundew bear numerous tentacles round their edges and on their upper surfaces. Each tentacle consists of a stalk with a globular head at its tip; this head is crowned by a glistening droplet of fluid, so that the leaves have the appearance of being bathed in dew. The tentacles round the margins of the leaves have relatively long stalks, while the stalks of those on the upper surface of the leaves are shorter. There may be as many as two hundred tentacles to a leaf.

Microscopical examination of a tentacle shows that a strand of xylem runs up the stalk, extending into the knob at its tip. Here the vascular strand is surrounded by three layers of cells. The innermost layer somewhat resembles an endodermis, the walls of the cells being thickened, while the two outer layers contain the red pigment that gives its colour to the leaf, and these layers are also concerned with secretion.

It has been suggested that the leaves have a certain fascination for insects, either through their colour or their glittering secretion or their smell, but this has not been proved. At any rate, any insect incautious enough to settle on a leaf of the sundew soon regrets it—though it has little time left to it for vain regrets. As soon as the insect touches a tentacle it becomes entangled in the fluid that the tentacles secrete, which is intensely viscid. The sundew does not rely on the viscid secretion alone to hold its prey, for, as soon as an insect has been caught the tentacle itself begins to bend over at its base, and other tentacles join in, so that the body of the insect is buried beneath a mass of sticky tentacles. Not only do the tentacles bend, but the whole leaf begins to fold along the centre of the blade, both holding the prey by closing up on it and bringing other tentacles into play.

Having trapped the insect, the sundew digests it. When an insect has been caught the tentacles begin to secrete a digestive enzyme allied to the pepsin that is found in the human stomach, together with an acid to provide the acid medium required for the working of the enzyme. The enzyme dissolves the soft parts of the prey, breaking down the proteins to simpler compounds that can be absorbed by the leaves. When all the digestible parts of the insect have been absorbed, the leaf opens again and is ready for another victim.

The bending of the tentacles of the sundew appears to be due to unequal growth, as in the haptotropic response of the tendril of a climbing plant to contact (Chapter 11). One presumes that a redistribution of auxin takes place in response to contact, though this has not been demonstrated. That the stimulus that evokes the reaction is a contact one is shown by the fact that the tendrils of the sundew will bend when touched with a hard neutral object such as a piece of glass. Like the tendrils of a climbing plant, the tentacles of the sundew do not respond to contact with fluids; raindrops leave them unaltered. The part of the tendril that is sensitive to touch is the globular head, while the actual bending takes place near the base of the stalk.

The tentacles are not only sensitive to touch, but also to chemical stimuli, for they will bend in response to treatment with certain substances in solution. Darwin was the first to point out that the substances capable of evoking a response in the tentacles all contained nitrogen. He tried *Drosera* leaves with various non-nitrogenous substances, including gum arabic, sugar, starch and alcohol, and in each case failed to produce any movement. When he tried nitrogenous products such as milk, albumen or meat infusion, he was successful in obtaining a response in every case. In particular, the tentacles were very sensitive to solutions of ammonium salts.

Undoubtedly the sundew benefits from the capture and digestion of prey; this was first demonstrated by Francis Darwin, son of the great Charles. Francis Darwin took six dishes full of sundew plants and separated them into two halves by placing a bar across each dish. Those on one side of the bar he fed with meat, while the others were left unfed. After a time, the plants fed on meat appeared more healthy and contained more chlorophyll and starch. The most dramatic improvement that resulted from meat feeding, however, was on the number of flowers produced by the two series of plants. When the flower stalks were cut in August, it was found that the meat-fed plants produced a hundred and sixty-five to every hundred produced by the control plants, and that the plants fed on meat produced two hundred and forty seeds to every hundred formed by the controls. It would seem, therefore, that the principal benefit from a diet of insects is increased reproductive capacity of the plants.

Sundew plants growing under natural conditions presumably benefit as much from the insects they catch as plants growing in the laboratory. What particular food elements they get from the insects is uncertain, but nitrogen is an answer that springs naturally to the mind, especially in view of the reaction of the tentacles to nitrogenous substances. In this connection, it is perhaps significant that the natural home of the sundew is in acid bogs and heaths, where the activity of nitrifying bacteria in the soil is curtailed by the acid conditions. What could be more natural than for a plant to supplement its diet with a few nitrogenous morsels of insect flesh, given the opportunity?

There are three British species of sundew, distinguished from one another by the shapes of their leaves. The genus *Drosera* is a large one, with ninety species, many of them tropical. All catch insects in the same way. Some of the tropical species are bigger than the British sundews; in *D. binata*, for instance, the leaves are long and forked. In the closely related *Drosophyllum lusitanicum*, a shrubby plant growing in Spain, Portugal and Morocco, insects are caught by sticky fluid secreted by glands on the leaves, there being no power of movement. The leaves of *D. lusitanicum* have been used as fly papers by Portuguese peasants.

The Venus's fly trap (*Dionaea muscipula*) is another member of the sundew family (Droseraceae) that catches insects in quite a different way. Dionaea is found in damp places, chiefly in North and South Carolina, and it has the distinction of being the first insectivorous plant to have its activities scientifically recognized, for in 1768 a London naturalist named Ellis sent a description of the plant to Carl Linnaeus, who described it as a 'miraculum naturae'. Linnaeus did not dream that the plant actually lived on insects, for he thought that the insects were captured accidentally and afterwards allowed to go free.

Dionaea traps insects by means of its leaves, and grows on wet moorlands, but there the resemblance to the sundew ends. The somewhat rounded leaves of *Dionaea* are in two halves which are capable of closing upon one another like a book; the edges of the leaves are ornamented with a number of marginal teeth. When the leaf closes, the teeth interlock with one another like the teeth of a gin trap (Fig. 51).

Near the middle of the upper surface of the blade of each half leaf are three delicate hairs. These are sensitive to touch. If a fly or a small worm touches one of these hairs while passing over the leaf, the leaf snaps shut, imprisoning the unfortunate victim inside the trap formed by its two halves. The surface of the leaf is studded with small red glands which secrete a fluid containing formic acid and a peptic enzyme. The soft parts of the insect are digested and the products of digestion absorbed by the plant. The process of digestion may take a week or more, during which time the leaf remains closed. When the prey has fully served its purpose the leaf opens again and is ready for more.

The digestive fluid is secreted by the glands on the leaf in considerable quantity. Darwin made an incision in a leaf that had closed over a large fly and observed that fluid continued to run from the cut during the whole nine days that it took the plant to digest the insect.

Aldrovanda vesiculosa is closely related to *Dionaea*. It is a rootless floating plant of sunny ponds in south and central Europe, India and

Australia. Its leaves are smaller than those of *Dionaea*, and are normally held in a half-open position, like the valves of a mussel. It has sensitive hairs on the surface of its leaves, which close in the same way as those of *Dionaea*. The prey consists of water fleas, insect larvae and diatoms. Although the leaves have glands on their surfaces, it is not known whether or not they secrete a digestive fluid.

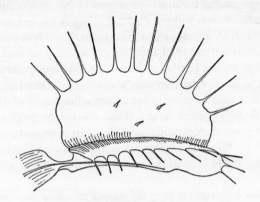

Fig. 51. A leaf of the Venus's fly-trap.
Note the sensitive hairs
in the middle of the leaf

Other carnivorous plants

All the insectivorous plants so far described belong to the sundew family. Also carnivorous, and not related to the sundews, is the bladderwort family (Lentibulariaceae), which has some two hundred and fifty species in all parts of the world. The simplest are the butterworts (*Pinguicula*), of which there are four species to be found in Britain.

The common butterwort (*P. vulgaris*) is found in bogs and on wet heaths all over the British Isles, though it is rare in the south. It grows well on mountain moorlands, ascending to a height of three thousand two hundred feet above sea level. The name *Pinguicula* means 'little fat one', a reference to the rosette of fleshy leaves which lies flat on the ground, with a single violet flower of great beauty rising from the centre. Insects are caught on the leaves by the sticky secretion from numerous stalked glands, while other glands, without stalks, secrete a digestive fluid.

Besides its peptic enzyme, the digestive fluid of the butterwort

223

appears to contain a substance allied to the rennin produced by the stomach of a calf. Rennin has the property of causing milk to curdle, and the butterwort is used in Lapland to curdle milk for making cheese.

The four British species of bladderwort (*Utricularia*) have a far more complex type of insect trap than the sticky leaves of the butterwort; they are among the most specialized of the carnivorous plants. They are water plants, without roots, floating freely in lakes, ponds, and ditches, mainly on mountain moorlands. The plant body apparently consists of a number of short shoots which bear finely divided leaves, though doubt has been thrown on this interpretation of its structure, and it may be that the 'shoots' are themselves modified leaves. For a month or more in the summer the flower of *Utricularia* is held aloft on an erect flower stalk, the bright golden blossom making the plant conspicuous as it floats in a mountain tarn. At other times in the year *Utricularia* is apt to escape notice, for the plants are small and float in the water with their leaves mainly submerged.

Utricularia captures its prey by means of small bladders, about one-tenth of an inch long in *U. vulgaris*, the commonest British species, which are carried in considerable numbers on the ends of the leaves, replacing some of the ultimate leaf segments. The mode of development of the bladder shows that it is a modified leaflet. The bladder is ovoid or pear-shaped, and is attached to the leaf by a short stalk that springs from one side.

The bladder is entered by a small valve, or trap door, which opens inwards; when closed, it rests against a thickened ledge around the mouth of the bladder. The interior of the bladder has on its surface a number of four-armed hairs which absorb water, as a result of which the hydraulic pressure inside the bladder becomes less than the pressure outside, as can be seen by the pinched appearance of the bladder walls.

On the outside of the trap door are sensory hairs: minute and sensitive to touch. If a small water creature accidentally touches one of these hairs, the 'spring' that keeps the trap door shut is momentarily relaxed. The trap door opens, pushed by the pressure of water outside, and water rushes into the bladder, carrying with it the hapless prey whose clumsiness started the reaction. The trap door shuts, and the animal is imprisoned.

Water fleas such as *Daphnia* and *Cyclops* form the bulk of the prey

224

collected by the bladderwort, but any creature small enough to enter the bladder but large enough to stimulate the sensory hairs may be caught. No digestive juices are secreted by the bladders; the plant waits for the visitors to die, and for bacteria to decompose their bodies. The soluble products of decomposition are then absorbed by the water-absorbing hairs, along with some of the water in the bladder. The bladder is then ready to receive the next meal.

Utricularia is a large genus, with over two hundred species, many of which are tropical. Many are land plants, and some are even epiphytes. All are carnivorous.

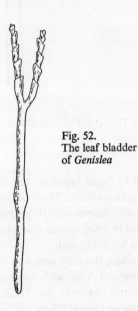

Fig. 52.
The leaf bladder
of *Genislea*

Genislea is an interesting genus of the bladderwort family that lives in tropical America and Africa. It is a land plant with curious pitcher-like leaves which lie closely appressed to the soil. The pitcher is formed by the lower part of the leaf, and opens by a forked neck with two twisted arms (Fig. 52). These arms have a lining of backward-directed hairs which allow small soil animals to enter the pitcher but prevent them from leaving. The interior of the pitcher is provided with glands, but whether these secrete digestive juices, or whether the plant waits for bacteria to do the work of digestion, we do not know.

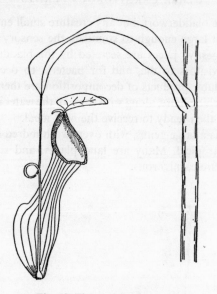

Fig. 53. The pitcher of *Nepenthes*

Pitcher plants

The pitcher plants belong to three families: the Nepenthaceae, the Sarraceniaceae, and the Cephalotaceae. Most of them are tropical or subtropical, but one species, *Sarracenia purpurea*, was introduced into a bog in central Ireland in 1906, where it has become thoroughly naturalized and appears to be doing well.

The pitcher plant *Nepenthes*, the only genus of its family, is found mainly on the islands of tropical Asia, with its greatest development in Borneo, though its range extends to Madagascar, northern Australia, Bengal and southern China. There are about sixty species, most of which are climbing shrubs growing in boggy places and supporting themselves by leaves modified to form tendrils. The remarkable pitchers in which they catch insects arise at the ends of the leaf tendrils.

The leaves of *Nepenthes* end in tendrils which after coiling once round a support curl downwards and then, rising once more, form pitchers at their tips. The pitchers vary from an inch to a foot in length; they are urn-shaped with a projecting flap at the top, formed from the tip of the leaf, not unlike the lid of a German beer mug (Fig. 53). It is said that this projecting lid prevents the pitchers from

226

becoming flooded with rain. The whole pitcher is often coloured with brilliant reds and purples, making it conspicuous and attractive to insects.

The pitchers of *Nepenthes* are beautifully designed insect traps. The opening of the pitcher is surrounded by a ribbed rim containing woody tissue which makes it rigid and keeps the mouth of the pitcher open. The inner wall is differentiated into two regions of quite distinct structure. The upper region consists of cells with waxy walls, forming a surface on which no insect can keep its feet; the waxy covering of the cell walls flakes off in tiny scales which clog even the adhesive papillae on the feet of ants. Any insect that ventures on to this treacherous surface can hardly fail to slip down into the pitcher.

The lower region is glandular, the multicellular glands secreting the watery fluid with which the pitcher is filled. It has a glistening appearance, owing to droplets of fluid secreted by the glands.

Nectar is produced by nectaries on the lower side of the lid, serving to attract insects to the pitchers. When the insects venture on to the rim of the pitcher a few steps are sufficient to carry them on to the slippery layer, from which there is no return. The liquid secreted by the glands in the pitcher is at first neutral in reaction, but as soon as an insect has been captured the reaction becomes acid, and a peptic enzyme appears in the fluid. This enzyme digests the soft parts of the bodies of the prey.

It is interesting that the pitchers of *Nepenthes* are always found, when cut open, to contain a varied flora and fauna, living in the fluid. These include blue-green algae, desmids, diatoms, Protozoa, nematodes, mites, flies and their larvae, and larvae of butterflies and moths. These lodgers appear to be perfectly healthy, and to be there by choice rather than victimization. One wonders how they protect themselves from the action of the digestive enzyme in the fluid. Possibly they have an anti-enzyme to render the enzyme ineffective. Such substances are known to exist.

The Sarraceniaceae is a family that includes three genera of pitcher plants, of which *Sarracenia* is the most important. *Sarracenia* grows in marshes in the warmer parts of Atlantic North America. It differs from *Nepenthes* in that the whole leaf is modified to form the pitcher, whereas in *Nepenthes* the pitcher is formed from only part of the leaf. The pitchers of *Sarracenia* are trumpet-shaped (Fig. 54) and stand erect. As in *Nepenthes*, the mouth of the pitcher is protected by an erect flap. Along the side of the pitcher is a flat expansion which not

Fig. 55. The pitcher of *Darlingtonia*

Fig. 54. The pitcher of *Sarracenia*

only increases the photosynthetic surface, but also provides a convenient ramp up which insects can crawl to the mouth of the pitcher.

The mouth of the pitcher in *Sarracenia* has a stiff rim which keeps it open. Nectaries abound on the lid, the rim and the outer surface of the pitcher, providing plenty of temptation for insect visitors. Nectar-seeking insects are lured to the inner edge of the rim, where the slippery zone begins. This consists of cells with smooth, waxy cuticles which overlap one another like the tiles on a roof. Below the slippery zone is a region that has long bristles, all of which point downwards into the depths of the pitcher; this zone also has nectaries between the hairs, so that insects are lured ever downwards. Finally, near the bottom of the pitcher there is a region where the surface is smooth, devoid of hairs, and without obvious glands.

The pitchers of *Sarracenia* usually, though not always, contain some fluid, but whether it is secreted by the nectar glands or by the undifferentiated epidermis of the lower part of the pitcher, or whether it is mainly rain water, is uncertain. It used to be thought that *Sarracenia* produced no digestive juices, but it is now known that an enzyme is present in the liquid in the pitchers. The digestible parts of

228

the bodies of the insects are broken down, and the products absorbed by the walls of the pitcher.

In some species of *Sarracenia* the fluid in the pitchers is alkaline instead of being acid. If the digestive enzyme functions in an alkaline medium rather than an acid one, it must be different from the peptic enzyme in *Nepenthes*.

Sarracenia appears to be less perfectly specialized for the insectivorous habit than *Nepenthes*. *Darlingtonia*, another member of the Sarraceniaceae that grows in California, has elaborate pitchers which have no enzymes for the digestion of insects, the work being done by bacteria in the fluid of the pitchers (Fig. 55).

The third genus in the Sarraceniaceae is *Heliamphora*, a small genus of pitcher plants of British Guiana and Venezuela.

Cephalotus follicularis, the only member of the Cephalotaceae, is a pitcher plant of the marshes at King George's Sound, Western Australia. Only the lower leaves form pitchers, the upper ones being flat and green. The pitchers of *Cephalotus* operate in much the same way as those of *Nepenthes*.

Predacious fungi

The carnivorous habit is not confined to the higher plants, for it is well developed in certain groups of fungi. Just as it has arisen several times in the course of evolution in the flowering plants—for the families that show the habit are not all closely related, and the means of capturing the prey are diverse—so in the fungi we find that predacious fungi, as the carnivorous fungi are called, crop up in widely separated orders, implying separate evolution.

The predacious fungi, like the insectivorous flowering plants, capture small animals alive and feed on them. The victims are mainly Protozoa and eelworms, though rotifers are sometimes taken. The Protozoa that fall victims to the predacious fungi are mostly small amoebae, or their relatives such as *Difflugia*, which have their bodies enclosed in hard outer shells. Rotifers are well known to the amateur microscopist; they are known as 'wheel animalcules' because of the circle of waving hairs, or cilia, round their mouths, the movement of which give the impression of a rotating wheel. Eelworms are microscopic nematode worms, or roundworms. They are extremely common, and they get their name from their eel-like appearance under the microscope, and especially from their rapid, threshing movements.

Among the lower fungi, the Zoopagales are outstanding as a carnivorous group, all but a few members of the order being predacious. The prey of the Zoopagales are mainly amoebae and other related Protozoa. The fungi are of simple construction, consisting of extremely fine branching threads or hyphae. When an amoeba comes into contact with a hypha of one of the Zoopagales it sticks, apparently held by a sticky fluid secreted by the hypha. The fungus then puts out a haustorium ('sucker'), which penetrates into the body of the animal, branching as it does so, and absorbs the body contents (Fig. 56).

At first the amoeba appears to be little perturbed by the intrusion of a foreign body into its interior, for its natural functions continue, as far as one can tell, unchecked. As the contents of its

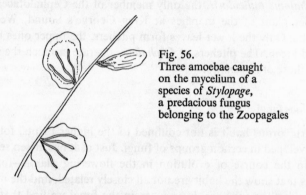

Fig. 56.
Three amoebae caught
on the mycelium of a
species of *Stylopage*,
a predacious fungus
belonging to the Zoopagales

body become progressively absorbed, however, the movements of its contractile vacuole—the pulsating vacuole which acts like the bilge pump of a ship, pumping excess water out of the animal—grow slower, and finally cease. The amoeba becomes more transparent as its protoplasm is absorbed by the fungus, and at some point in this stage—it is hard to say exactly when—it may be presumed to be dead. The fungus continues its work of absorption until only the shrivelled outer 'skin' or pellicle of the amoeba is left attached to the hypha.

Although most of the Zoopagales capture Protozoa, there are a few, more robust, that make eelworms their prey. An eelworm is a larger animal than an amoeba, and far more active; a fungus going eelworm hunting is not unlike a man catching sharks with rod and line. The hyphae of the eelworm hunters are stouter than those of most of the Zoopagales that content themselves with Protozoa. The means of capture, however, is the same. The eelworms are caught by a sticky secretion of the hypha which holds them as relentlessly as bird-lime entangles a fluttering bird (Fig. 57).

An eelworm is a more sporting prey than a stolid amoeba. From the moment it is caught it fights for freedom like a hooked tarpon, threshing its body this way and that, and pulling the hypha that holds it until a break seems imminent. The fungus is as tough as the eelworm, however; it holds on like a bulldog until the eelworm, exhausted with struggling, becomes quiescent. Then, from the part of the hypha in contact with the prey, more hyphae grow out into the

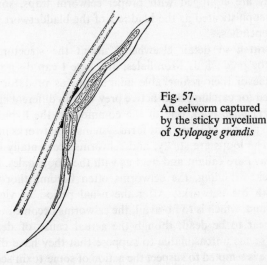

Fig. 57.
An eelworm captured
by the sticky mycelium
of *Stylopage grandis*

body of the eelworm and consume its contents. In less than twenty-four hours nothing but the integument of the eelworm is left, filled with empty hyphae whose contents have been passed back to the fungus.

The Zoopagales appear to be obligate predators, for nobody has yet succeeded in growing them on an artificial culture medium without their prey. They are extremely common in soil, dung and rotting vegetation generally; wherever their victims are found, the Zoopagales are likely to be. But in spite of their being so common they were discovered only fairly recently, for until Drechsler published the first of a long series of descriptions of them in America, nobody dreamed of their existence. Drechsler described the first of the Zoopagales in 1935, and during the last thirty years has added many more to his list. Thirty years ago the Zoopagales were a minor botanical curiosity; today they are a major group of fungi.

To see the sport of eelworm hunting at its best we must turn to another group of fungi, far removed from the Zoopagales both in structure and in relationship. These are the predacious Moniliales. The Zoopagales have no morphological adaptation for catching prey; most of them do not need any, for the Protozoa are easy game, and only a few species aim higher. The predacious Moniliales, on the other hand, make eelworm-catching their main activity, only one or two species accepting humbler prey. It is not surprising, therefore, to find that they are equipped with proper eelworm traps, some of which are as sophisticated as the bladders of the bladderwort or the pitchers of Nepenthes.

I have written in detail elsewhere* about the structure and activities of the predacious Moniliales, and here I can do no more give an outline of their remarkable adaptations as predators. The means adopted for catching their active prey vary in different species. In *Arthrobotrys oligospora*, one of the commonest, the hyphae are provided with a series of loops that form systems of networks in three dimensions. The loops are sticky, and eelworms accidentally brushing against them are caught and held as with the Zoopagales. In the course of their wriggling, the eelworms often become thoroughly entangled with the networks. After the usual period of vigorous threshing around, which is to no avail, the eelworms become motionless, and appear to be dead, though the actual cause of death is uncertain. It seems unreasonable to suppose that they have died of fright, and one is tempted to suspect the action of some toxin secreted by the fungus. Some workers have claimed to have isolated such a toxin, but their work needs confirmation before it can be accepted unreservedly.

As soon as the eelworm is moribund, a minute papilla put out by the fungal network penetrates its integument and swells inside the animal to form a globular infection bulb. Hyphae grow out from this and fill the body of the eelworm, absorbing its contents. When all is consumed, the protoplasmic contents of the hyphae pass back into the networks for the nourishment of the fungus, leaving the integument of the eelworm, filled with empty hyphae, attached to the network.

The voracity of *Arthrobotrys oligospora* has to be seen to be believed. The fungus can virtually exterminate the eelworms in a

*The Friendly Fungi (Faber and Faber, 1957).

Petri dish culture in a matter of a few days, if it really gets going in earnest.

Other variations on the sticky networks are found in the many different species that make up the predacious Moniliales. Some, such as *Monacrosporium cionopagum*, have short lateral branches on their hyphae which are sticky and act in the same way as the networks of *Arthrobotrys*. *M. ellipsosporum* and others have small sticky knobs attached to their hyphae by short stalks. Details of the capture of eelworms are similar to those described for *Arthrobotrys*.

Not all the predacious Moniliales rely on sticky traps for catching eelworms. In *Dactylaria candida* we find a mechanical form of trap. The hyphae have numerous small rings, each composed of three curved cells, attached to them by short stalks. Should a wandering eelworm pass its body into a ring it sticks fast, for the diameter of a ring is slightly less than the diameter of an eelworm's body. Lacking the sense to withdraw while there is yet time, the foolish creature tries to bullock its way through, with the result that it becomes inescapably wedged into the ring. Events then follow their usual course, the body contents of the eelworm being absorbed by hyphae that grow out of the ring.

The rings of *Dactylaria candida* are not sticky, nor have they any power of contraction about the body of the eelworm, which is caught purely by its own efforts to slide through the ring. In some of the predacious Moniliales, however, the rings are far more deadly; as soon as an eelworm enters a ring it closes, holding the eelworm as tightly as a rabbit caught in a snare.

Dactylaria gracilis is one of the predacious fungi equipped with constricting rings, very like those of the non-constricting type found in *D. candida*, except that they are a little stouter in construction, and their stalks are shorter and thicker. The three cells of the ring are sensitive to touch on their inner sides, and when an eelworm wanders into a ring the friction of its body sets the trap in operation; the three cells of the ring suddenly swell inwards to about three times their previous size. This occludes the opening in the ring, so that the body of the eelworm is tightly constricted beyond hope of escape. Capture of an eelworm is followed by the growth of absorptive hyphae into its body, as before. The operation of the constricting rings is amazingly quick: after a lag phase of a few seconds, the actual swelling is accomplished in one-tenth of a second. The eelworm does not have a chance of escape, once the mechanism is triggered off. A great deal of

233

work has been done on the mechanism of the constricting rings, but nobody yet knows how it works.

The predacious Moniliales differ from the Zoopagales in that they can easily be isolated and cultured in the laboratory, though they do not usually form their eelworm traps unless eelworms are also present in the culture. The ease with which they can be cultivated has led many workers, in England, America, France and Russia, to attempt their use for the biological control of plant parasitic eelworms, among the worst enemies of our crops.* Some slight success has been achieved on the experimental scale, but there is a great deal more to be done before they are likely to save part of the £2,000,000 that the potato root eelworm alone costs this country every year.

The predacious Moniliales are as abundant as the Zoopagales, if not more so. They occur in soil, in the dung of animals, in rotting vegetable matter—in fact wherever eelworms flourish you will almost certainly find the predacious Moniliales preying on them.

The carnivorous plants are a remarkable and interesting ecological group, and a group that defeats the imagination when one speculates on how they have evolved. Up to the moment they are economically useless—if we except the use of the leaves of *Drosophyllum lusitanicum* as fly papers by the Portuguese—but this does not detract from their interest. Besides, one never knows. Perhaps one day somebody will breed a tomato plant that eats its own white fly, or a really efficient aphis or slug eater may be found and brought back from the forests of the Amazon to the British greenhouse. I do not think, though, that it will be in our time!

*See *The Friendly Fungi*.

14 · Plants in partnership

PLANTS IN PARTNERSHIP

The fungi are generally regarded as the arch-enemies of the higher plants; probably with justification, for more plant diseases result from the attacks of fungi than from any other cause. Their relationship is not always that of a predator and prey, for there are instances where they enter into partnership for mutual benefit.

If some of the shallow-feeding roots of the beech are dug up, it will be found that near their tips there are many short branches which are responsible for most of the uptake of water and minerals from the soil (Fig. 58). If a cross-section is cut of one of these rootlets, it will be seen to be closely covered by a weft of thread-like fungal hyphae, to a depth of from 20 to 30μ (the micron, represented by the Greek

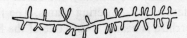

Fig. 58. Mycorrhizal roots of the beech

letter μ, is the microscopist's unit of measurement, 1μ being equal to one-thousandth of a millimetre).

Here and there, hyphae branch off from the mantle and run into the soil. The hyphae also penetrate the root to some extent, passing in between the cortical cells and forming what is known as the 'Hartig net'. These internal hyphae do not penetrate very far into the root; normally they are confined to the outer cells of the cortex, and they never extend inwards beyond the endodermis.

Symbiotic associations

It might at first be thought that the fungus is parasitic on the beech tree, but this is not the case. The tree is not in any way harmed by the

235

fungus on its roots; in fact, it appears to benefit from the association. The two organisms are in partnership, each gaining some advantage from living with the other. For such associations as this, where two organisms share the same body for mutual benefit, biologists have coined the term 'symbiosis'. There are many such symbioses between plant and animal, or plant and plant, and this particular association between a fungus and the roots of a higher plant is called mycorrhiza (literally, 'fungus root').

A similar association occurs between the roots of the pine and a fungus. As in the beech, the fungus invests the short, stumpy branches that are formed near the ends of the lateral branches from the main root. In the pine these mycorrhizal branches tend to fork and their ends are characteristically swollen, forming what are called coralloid mycorrhizas, from their fancied resemblance to a coral.

Where a mycorrhizal association occurs, the formation of root hairs near the tip of the root is largely or completely suppressed, and the threads of the fungus take their place. This they do with considerable adroitness; it is thought, in fact, that they are more efficient absorbing organs than the original root hairs. Mycorrhizas of this kind are found on the roots of many forest trees. They are common in the Pinaceae (pine, fir, larch, etc.) among the conifers, and also occur in the birch family (Betulaceae), the beech family (Fagaceae), and some other families of flowering plants. The fungal partner in this type of mycorrhiza is nearly always one of the Basidiomycetes— the group of fungi that includes the mushrooms and toadstools. Many of the toadstools that are so common in woodlands are the fructifications of the fungi that are associated with the roots of the trees. In some cases, the fungus is quite specific to a certain species of tree; *Boletus elegans* is always associated with the larch (*Larix decidua*). In other instances, a particular fungus may form mycorrhizas with a great variety of different trees. The record seems to be held, at the moment, by *Cenococcum graniforme*, which is known to form mycorrhizas with pine, fir, spruce, Douglas fir, hemlock spruce, larch, juniper, oak, beech, birch, hazel, hickory, lime, poplar, willow, and alder. It is not for many mycorrhizal fungi that we can produce such a comprehensive list as this, and most fall somewhere between the choosiness of *Boletus elegans* and the catholicity of *Cenococcum*.

It is uncertain to what extent mycorrhizal fungi are dependent on their hosts for existence. Some species of mycorrhiza formers, such as *Boletus bovinus*, seem to be able to get along on their own without any association with the roots of trees, but for most species it is doubtful whether the fungus is able to live independently. As for the trees, it would seem that some trees which normally form mycorrhizas are capable of growing quite well without their fungal partner, at any rate when they are planted in a rich soil; in poor soil, the need for the fungus is greater. The experiments of Rayner and Neilson Jones showed that on the poor mineral soil of Wareham Heath pines would not make good growth unless they were infected with their mycorrhizal fungi.

Although much work has been done on problems relating to mycorrhiza during the past two decades, we do not yet know what it is that each partner gains from the partnership. An early theory that has received much publicity was that the fungus was able to break down complex organic compounds in the detritus of the forest floor, and thus make them available for the tree. Attractive though this theory is, it must be viewed with grave doubt except in a few instances. For one thing, the known properties of the fungi themselves, as determined by experiment, make any such phenomenon extremely unlikely. Most mycorrhizal fungi are unable to obtain their carbon from complex organic carbon compounds, requiring a supply of sugar as their carbon source. It seems far more likely that the mycorrhizal habit increases the absorbing power of the roots by the greater spread of the fungal hyphae compared with normal root hairs, and also possibly by the greater efficiency of fungal hyphae in absorbing minerals both from soil and from the mass of decaying litter on the forest floor. Phosphates in particular have been mentioned as being gathered readily by the hyphae.

Assuming that the tree benefits in this way, we have still to solve the problem of how the materials absorbed by the fungus are passed on to the host plant. There is no theoretical reason why materials passing along the hyphae should not be passed into the root cells via the Hartig net, but it has not been demonstrated that this is so. The amount of soluble material produced by the breaking down of organic litter varies greatly at different times. It may be that the fungus absorbs phosphates or other minerals in times of plenty, passing them on to the host when they are scarce.

If we do not know how the tree benefits from the mycorrhizal state, we are equally ignorant of the advantages to the fungus. The most plausible theory is that the tree supplies the fungus with ready-elaborated organic compounds; sugar seems to be the most obvious thing that the fungus might lack. In this connection it is perhaps significant that the establishment of mycorrhizal associations seems to depend on the nutritional status of the tree; seedlings, for instance, are not attacked—if that is the right word—by mycorrhizal fungi until after the first leaves have opened and the young tree is beginning to supply itself with carbohydrates by photosynthesis.

Ectotrophic and endotrophic mycorrhizas

The mycorrhizas formed by forest trees are called 'ectotrophic' mycorrhizas, because the fungus is mainly on the outside of the root. There is another kind, the 'endotrophic' mycorrhiza, in which the fungus penetrates the host tissues to a much greater extent, its hyphae entering the host cells and setting up a more intimate connection between the two plants. The former is not as widespread as the endotrophic type; the more work that is done on this, the more nearly universal it appears to be. Very few families of plants are known to be without any members that show endotrophic mycorrhiza.

It is perhaps a pity that the term mycorrhiza came to be used for both ectotrophic and endotrophic types, for they have little in common except in extreme cases. Moreover, there are undoubtedly several different kinds of endotrophic mycorrhiza, some of them showing great divergencies one from another. The plants of the heath family (Ericaceae) have an endotrophic mycorrhiza that is in some ways intermediate between the ectotrophic mycorrhizas of forest trees and the highly specialized endotrophic mycorrhiza of the orchids. The roots of such plants as the heathers (*Erica*), the ling, and the whortleberries (*Vaccinium*) are extremely fine, with a narrow cortex and a central vascular region. They are enclosed in a loose mantle of fungal hyphae which penetrates the cortex, the hyphae actually entering the cortical cells to form compact hyphal masses within them. The hyphae which thus enter the cortical cells may later be digested by the host.

The stages of infection of a young root follow a definite pattern. Infection is lightest during the spring when the roots are making rapid growth, and hyphae that do succeed in invading the cortex are quickly digested. Later in the year infection becomes heavier, reaching a climax in the early autumn.

Both in the formation of an external mantle of hyphae and in the seasonal variation in the incidence of infection, the mycorrhiza of the Ericaceae resembles the ectotrophic mycorrhiza of forest trees. The ericaceous mycorrhiza does differ from that of the trees in the extent to which the cortical cells are penetrated by the fungus. In tree mycorrhiza, penetration into the cortical cells is rare as the fungus of the Hartig net spreads *between* the outer cells of the cortex rather than entering them. In the ericaceous type of mycorrhiza the fungus enters the cortical cells freely; this, together with the ultimate digestion of the hyphae by the host cells, is a fundamental difference between endotrophic and ectotrophic mycorrhiza.

Rayner and other workers on mycorrhiza have reported the presence of the fungus in other parts of the plant beyond the roots: in stem, leaves, and even flowers and fruit. The fungus present in the aerial parts of the plant was said to be in an 'attenuated' condition, the hyphae being finer than the root fungus and not staining deeply. This led to the belief that in the Ericaceae mycorrhizal infection was systemic: that is, the whole plant was normally infected.

This claim has been investigated by a number of later researchers with entirely negative results, and one must conclude in the light of recent work that systemic infection is not a normal part of the ericaceous mycorrhiza, which ordinarily affects the roots and the roots only. Where the attenuated hyphae seen by Rayner and others came from is uncertain; they may have belonged to quite a different fungus, or possibly a systemic infection does occur in certain abnormal cases, as, for instance, when the resistance of the host to the fungus happens to have been lowered through some cause. The balance between symbiosis and parasitism in endotrophic mycorrhiza is a fine one, and there is no reason to suppose that the mycorrhizal fungus could not become a straightforward parasite, given suitable conditions.

The identity of the fungus causing mycorrhiza in the Ericaceae has also given rise to considerable controversy. It was formerly believed to be a Basidiomycete, *Phoma radicis*. This fungus was isolated from the roots of various species of Ericaceae, including the cross-leaved heath (*Erica tetralix*), by Ternetz in 1907. It has been isolated from seeds of the ling on several occasions and was again obtained from roots, this time of *Vaccinium oxycoccus*, by Rayner and Levisohn, in 1940.

It is not sufficient to isolate the fungus from the roots of members of the Ericaceae and discover its identity; in order to prove that this particular fungus is the cause of mycorrhiza, it is necesssary to produce mycorrhizas artificially by inoculating seedlings, grown under sterile conditions, with the fungus. Up to now, *Phoma radicis* has failed to pass this test. We cannot, therefore, accept the hypothesis that *Phoma radicis* is the cause of mycorrhiza in the Ericaceae. Its isolation from seeds in any case loses much of its validity, since the theory of systemic infection is no longer widely held. We are left with its isolation from roots by Ternetz, sixty years ago, and one more recent isolation by Rayner and Levisohn. It could well be that the fungus obtained in these instances was not the cause of the mycorrhiza, but merely associated with it, possibly as a simple parasite.

If the fungus in question is not *Phoma radicis*, what is it? The answer to this question may be provided by the researches of Doak, who in 1928 isolated some sterile fungal hyphae from the roots of *Vaccinium corymbosum* and *V. pennsylvanicum*. As this mycelium produced no spores it could not be identified, but Doak was able to show that it would produce typical mycorrhizas when back-inoculated into sterile seedlings. The success of this inoculation experiment immediately gave Doak's fungus a stronger claim than *Phoma radicis* to be the causative organism of mycorrhizas.

This initial work by Doak was followed up by others, with similar results. Unfortunately, the hyphae isolated from the roots by these workers uniformly failed to form spores, so that identification was impossible; but one thing was certain: the fungus was not *Phoma radicis*. We do not yet know the identity of the fungus causing mycorrhizas in the Ericaceae, but we can say that it is unlikely to be *Phoma radicis*.

Nitrogen fixation

Yet another controversy that has arisen over mycorrhiza formation in the Ericaceae relates to whether the mycorrhizal fungi can or cannot fix atmospheric nitrogen. Some of the early work suggested that the fungus, by causing nitrogen from the atmosphere to combine with something contained in its hyphae, could render it available for the nourishment of the host plant. These findings were generally accepted without proper criticism and confirmation, and so there arose a kind of fairy tale in which the 'nitrogen fixation' in the Ericaceae was assumed. Modern work has shown the falsity of such ideas, and nobody today seriously believes in the power of the Ericaceae to fix nitrogen through the activities of their mycorrhizal fungi. As far as we are aware, no fungus of any kind can fix atmospheric nitrogen.

Although we are reasonably certain now that mycorrhizal fungi in the Ericaceae do not fix nitrogen, we are a long way from knowing what it is that they do to benefit the host. It has been observed that the early growth of seedlings of the Ericaceae is stimulated by the presence of fungi, but this appears to apply to almost any fungus, not strictly to those that form mycorrhizas. The resemblances between the mycorrhiza in the Ericaceae and that in forest trees suggests a similar kind of relationship; possibly mycorrhizal roots are better for absorbing minerals from the soil than uninfected roots. But positive evidence is lacking. There is also the possibility that the fungus may be capable of breaking down organic carbon compounds in the soil and making the products available to the host. We shall not find the answer until we know a great deal more about the mycorrhizas and the fungi that cause them.

What the fungus gains from the host is also at present a mystery. A supply of sugars in an obvious suggestion, but the right answers in biology are not always the obvious ones. Much more work will have to be done before we know why the fungus enters into this curious association.

Mycorrhizas of orchids

In the orchids we find mycorrhizas of a different type. Here there is no investing mantle of fungal hyphae round the roots; the hyphae are almost entirely inside the root, although some may leave the roots and connect with a mycelium in the soil. As we have seen, the orchids are a highly specialized family of individualists, and one of their characteristics is the production of enormous numbers of very small seeds. The seed of an orchid is a simple structure. The fertilized egg cell gives rise to an embryo which is small and unspecialized and may in exceptional cases consist of as few as eight cells, the number being seldom more than a hundred. There are no cotyledons; the embryo is simply an ovoid or fusiform mass of cells, with a growing point at one end.

When an orchid seed germinates it swells, the seed coat is ruptured, and a small amount of growth takes place, using the food material laid down in the seed. In their natural habitat further development is dependent on infection by the appropriate mycorrhizal fungus, for

241

the next stage in the growth of the seedling is saprophytic in all orchids. Fed by organic matter which it absorbs from outside, the embryo develops into a protocorm (see page 158). From this the various organs differentiate, after which chlorophyll is developed in the leaves and the young plant gradually begins to feed itself by photosynthesis.

An important point in this mode of development, and one that has a considerable bearing on mycorrhiza in orchids, is that *all* orchids pass through a saprophytic stage during their development. In most of them the saprophytic phase is short, the plant becoming green as the first leaves acquire chlorophyll, but some orchids never lose their saprophytic habit. We have seen how the bird's nest orchid (*Neottia nidus-avis*) grows up free from chlorophyll, remaining a saprophyte for the whole of its life.

The young protocorm is the first part of the orchid seedling to be infected by mycorrhizal fungus, and one would think that the roots would be infected from the protocorm; but they are not. The roots, like the protocorm, have to be infected from the soil. The protocorm is heavily infected, and so are the young roots; the tubers are free from fungus. As new roots grow out of the tubers in the following year they become infected with hyphae from the soil. This process of re-infection normally takes place every year.

The fungal hyphae that invade the cortex enter the cortical cells, forming coils of hyphae that at first appear to be perfectly healthy. Later the hyphae are digested by the host cells, forming a structureless mass. The digestion of hyphae may take place anywhere in the cortex, but is frequently confined to a definite digestive layer.

The fungi that cause mycorrhizas in orchids are fairly easy to isolate, and many of them have been identified. They seem to fall into two groups. Those isolated from green orchids belong mainly to the genus *Rhizoctonia*, a genus which, from its failure to undergo sexual reproduction, is placed among the Fungi Imperfecti—those fungi in which sex is unknown, and in which reproduction takes place solely by means of asexual spores. The fungi isolated from the roots of saprophytic orchids, on the other hand, are mainly Basidiomycetes such as *Xerotus javanicus* and *Marasmius coniatus*. The honey agaric (*Armillaria mellea*) is also a mycorrhizal fungus with certain orchids, such as species of *Gastrodia* and *Galeola*.

It has been shown that there is a certain amount of specificity in the relations between certain species of *Rhizoctonia* and certain species of orchids, though not all species of *Rhizoctonia* are the same in this respect. Early workers formed the opinion that the various species and strains of *Rhizoctonia* were highly host-specific, a particular strain of fungus forming mycorrhizal associations with a particular species of orchid, or a group of related species. It is now known that this is not so, though in many cases a certain strain of *Rhizoctonia* will infect one species of orchid more readily than it will attack others. We do not know enough yet to generalize about this.

Research on the physiology of orchid fungi has shown that they differ from the fungi of ectotrophic mycorrhizas in that they are able to make use of complex sources of carbon. This is important in view of the part they are thought to play in the saprophytic nutrition of orchids. If they had been unable, like the fungi from forest trees, to use complex carbon sources and in need of sugars as their carbon source, they could have played little part in the saprophytic lives of the orchids. As it is, it seems reasonable to suppose that the breakdown of complex organic matter may be their principal, if not their sole, contribution to the mycorrhizal partnership.

Assuming that the orchid fungi contribute the breakdown products of complex carbon compounds to their hosts, it remains to be decided how these products are passed on from the fungus to the orchid. One way of course would be by the digestion of the fungal hyphae in the root cells. There can be no doubt that the orchid does gain material in this way, but this is probably not the only method in which it is nourished by the fungus. It is at least reasonable to suppose that translocation of materials can occur along the fungal hyphae, and that some of the material translocated can be passed on, through the walls of the hyphae, to the cortical cells of the orchid. This would bring the orchid mycorrhiza into line with what is suspected to happen in the ectotrophic mycorrhiza of forest trees. There is no proof of this, but it does have the merit of probability.

What the fungus gets out of its association with orchid roots has not yet been established. The relation between the two plants is a somewhat delicate one; at times the fungus may become parasitic on the orchid and destroy it, while at others it may fail to make the initial infection of the seedling. The digestion of fungal hyphae in the cortical cells of the orchid is an obvious defence mechanism; if the activities of the fungus in its tissues begin to embarrass the orchid, it responds by digesting the fungus. It seems that what we have here is a sort of balanced parasitism; as long as the balance is maintained both partners thrive, but if one partner gets on top the result is fatal for the other. The answer to the question of what the fungus gains from the partnership would seem to be this: it gains that which is gained by any internal parasite, with the proviso that its parasitic activities are kept under control by the host.

Many other plants have mycorrhizal associations with fungi in a similar way to the orchids. In particular, we find saprophytes in the Triuridaceae and the Burmanniaceae with mycorrhizas similar to those found in orchids, and species of *Rhizoctonia* have been isolated from them. Nor are the mycorrhizas confined to the flowering plants, for the saprophytic gametophyte of *Psilotum*, a primitive relation of the ferns, also has the habit. Many liverworts are also known to be mycorrhizal.

Lower fungi in mycorrhizal association

The fungi concerned in all the mycorrhizas that I have described have 'septate' hyphae—that is, the hyphae are regularly divided by cross partitions, or septa, into separate cells. The possession of septate hyphae is a mark of the higher fungi, the lower fungi having none; the hyphae form long tubes, lined with protoplasm and containing a continuous central vacuole. Hyphae of this kind are also found in mycorrhizal association with higher plants; in fact, the mycorrhizas formed by the lower fungi are far more numerous than any other type. Recent work has shown that they are far more common than we had formerly supposed. Nearly all families of plants that have been investigated contain at least some genera with this type of mycorrhiza.

Affected plants show non-septate hyphae, often of large diameter, running through the root cortex and often producing branched structures like small bushes in the cells. These bush-like branch systems are known as 'arbuscles'. The hyphae may also form swellings, called 'vesicles', either on their ends or here and there along their lengths.

Nothing is known about the relations between host and fungus in these 'vesicular-arbuscular' mycorrhizas, but as the host plant appears perfectly healthy these are presumed to be symbiotic partnerships rather than parasitic infections. They are known to occur in a number of crop plants of great economic importance, including the cereals wheat, barley, oats, rye and maize. They are also found in clover and many other plants of the clover family.

The fungus concerned in this type of mycorrhiza has been given the name *Rhizophagus*, though this is a matter of convenience more than anything else, for we do not know whether the same fungus is concerned in all instances of mycorrhiza, or whether, as seems more probable, we are dealing with a whole group of fungi. There is good evidence that some at least of these mycorrhizas are caused by the fungus *Endogone*, common in soils. There are also indications that a species of *Pythium* may also be concerned. We eagerly await more work on these interesting fungi and on the mycorrhizas they cause.

An interesting point about mycorrhizas produced by *Rhizophagus* is their antiquity, for their hyphae, complete with vesicles, are found in fossils from as far back as the Devonian period, more than two hundred and seventy million years ago. The earliest fossil record is that of *Palaeomyces asteroxyli* which was found in the cortex of *Asteroxylon mackiei*, one of the earliest land plants.

Algae in partnerships

Besides forming mycorrhizal associations with fungi, many of the higher plants form symbiotic partnerships with algae. The blue-green alga *Anabaena*, for instance, is found in association with the tiny water fern *Azolla*, in the apogeotropic roots of cycads, and in a number of other plants. *Nostoc*, another blue-green alga, associates with the liverwort *Anthoceros*, with the cycads, and with certain flowering plants. There is evidence that in some cases at least the alga is able to pay its way by fixing atmospheric nitrogen.

The element nitrogen takes an especially prominent place among the chemical elements that are necessary to support life, for it is needed in larger quantities than most; and it is often scarce. About four-fifths of the air that surrounds the earth consists of nitrogen gas; this means that above every acre of the earth's surface about thirty thousand tons of nitrogen hang suspended. Unfortunately, very few organisms are able to make any use of this inexhaustible reservoir of nitrogen, for nitrogen gas is chemically inert and it is difficult to make it combine with other things.

Certain soil micro-organisms are nitrogen-fixers. They can absorb nitrogen gas from the air and combine it with other molecules to form proteins and other nitrogen compounds. Some of these nitrogen-fixers are blue-green algae, others are bacteria. Of the latter, *Azoto-bacter chroococcum* and *Clostridium pasteurianum* are the best-known free-living forms, and are responsible for replacing a great deal of the nitrogen wastage from the soil that occurs from a variety of causes, including the removal of nitrogen in harvested crops.

Nitrogen-fixing bacteria

More important than these free-living nitrogen-fixers are the nitrogen-fixing bacteria that enter into symbiotic partnerships with higher plants. The most familiar are the species of *Rhizobium* that inhabit the nodules on the roots of plants belonging to the order Leguminosae, and especially the family Papilionaceae, (characterized by butterfly-shaped flowers) to which peas, beans, and clover belong.

The *Rhizobium* bacteria in the soil are small and cylindrical, but

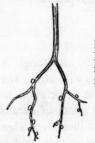

Fig. 59.
Rootlets of the
runner bean, showing the
root nodules

when they are living in the root nodules of their hosts they often form curiously-shaped branched cells known as 'bacteroids'. Seedlings of leguminous plants are infected early in their life by nodule-forming bacteria that enter their roots from the soil, infection taking place by way of the root hairs. Once inside the root hair the bacteria form an infection thread consisting of massed bacterial cells within a sheath of cellulose, hemicelluloses and pectins formed by the infected plant cell. The infection thread grows and branches many times, passing from cell to cell in the cortex of the root.

Infection of the root is followed by the formation of the root nodules—small globular swellings on the outside of the root (Fig. 59). In the pea and bean the nodule grows out of the cortex of the root, while in clover and lucerne (alfalfa) it is formed from the 'pericycle'—the outer part of the central cylinder of the root that houses the water-conducting tissues.

246

When the nodule is fully formed it consists of two layers of cells: a thin outer layer or cortex, and a central part or medulla. The whole nodule is covered by a thin layer of cells that is part of the cortex of the root. It is connected with the central water and food conducting part of the root by strands of vascular tissue (Fig. 60).

The bacteria, found in the medulla of the nodule, are very numerous; each cell of the medulla contains several thousand of them, and the whole nodule may house from one million to one thousand million of these tiny guests. If the nodule is a young one most of them will be alive and working, but in older nodules ninety-nine per cent or more will be dead.

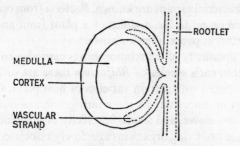

Fig. 60. Diagram of a section through
a root nodule

That the association between the leguminous plant and the bacteria is a symbiotic one cannot be doubted. Plants grown in nitrogen-deficient soil without their bacteria are pale, and make poor growth; they show all the symptoms of nitrogen starvation. In the presence of their bacteria, on the other hand, the plants make good growth even where the nitrogen content of the soil is low.

The return gained by the bacteria for the nitrogen that they give to the leguminous plant is less obvious, but it seems fairly certain that they are supplied with sugars by their hosts.

This symbiosis between *Rhizobium* and members of the Papilionaceae benefits the soil as well as the partners. Nitrogenous compounds are excreted into the soil by the growing plant, and on the death of the plant the soil also profits from the nitrogen compounds set free when its body decays. If the plant is a crop that is harvested, the soil still gains in nitrogen from the stubble that is ploughed in. That leguminous crops enriched the soil was known long before the existence of this symbiosis was suspected, and the knowledge was incorporated in

247

the famous Norfolk four-course rotation of crops, where wheat, a nitrogen-hungry crop, followed clover.

There are various species of *Rhizobium*, each of which can live in symbiosis with a particular group of leguminous plants. *R. leguminosarum*, for instance, is found in the roots of peas, broad beans, vetches, and lentils; *R. meliloti* occurs in lucerne and certain kinds of clover; *R. phaseoli* is associated with French and runner beans; *R. trifolii* partners red, white, and crimson clovers; *R. lupini* is symbiotic with lupins; and *R. japonicum* is the bacterium from the soya bean. If a partnership is to be set up, the right species of bacterium for the particular crop must be present in the soil. At least twenty-one of these plant-bacterial groups are known. Bacteria from one group may sometimes be found in the nodules of a plant from another group, but such partnerships are often unsatisfactory.

The host-specificity of the relationship is carried even farther than this, for within each species of *Rhizobium* there are different strains that are associated with certain varieties of host plant. One strain of *R. trifolii*, for instance, may fix more nitrogen in association with red than with white clover, and so on. Specificity of this kind is not at all unusual where two living organisms are closely associated.

The association between leguminous crops and *Rhizobium* has important agricultural implications, of which one, the position of clover in the rotation of crops, has already been mentioned. Another is the use of leguminous plants such as lucerne for green manuring. Under good conditions, a leguminous crop may fix upwards of two hundredweight of nitrogen per acre, and one hundredweight per acre is a fair average. The wise farmer cuts down his fertilizer bills by growing his own.

If the soil contains large quantities of nitrates the partnership between *Rhizobium* and its host breaks down. The plant absorbs nitrates through its roots in the usual way, and the bacteria in the root nodules fix little or no nitrogen as they can get what they want the easy way, from the soil. Under these conditions, the bacteria cease to be symbionts and become ordinary parasites on their leguminous hosts.

Under normal conditions, leguminous plants can pick up *Rhizobium* from the soil, but if the right strain does not happen to be present, or if it is present in insufficient quantities, soil inoculation may be needed. This can be done in a number of ways. The traditional method was to take top soil from another field on which a leguminous crop had been recently grown, and scatter it over the surface of the new field at the rate of four hundredweights per acre. Alternatively, soil from a field in which a suitable strain of *Rhizobium* was known to occur could be steeped in water, and the seed to be sown soaked in the water and then dried before sowing. A more modern technique is to use a pure culture of *Rhizobium* to treat the seed.

Lichens as partnerships

Finally, no account of plant partnerships would be complete without mentioning the lichens. These curious plants have been recognized as entities since the time of the Ancient Greeks, if not before, and it was the Greeks who gave us the word 'lichen', meaning 'scaly'. As a description of their dry, greenish-grey thalli it could hardly be more apt.

A lichen is composed of an alga and a fungus which have set up in partnership. Most of the lichen thallus is contributed by the fungus and is constructed of interwoven hyphae. In most lichens the unicellular algae which make up the partnership are arranged in a definite zone within the lichen, just beneath the surface, the alga concerned usually being a member of the green algae or Chlorophyta. Some lichens have the algae fairly uniformly spread throughout their bodies, in which case the alga is usually one of the blue-green algae or Cyanophyta. The fungal member of the partnership is nearly always one of the great group of fungi known as the Ascomycetes. Some lichens, however, are formed from fungi belonging to the Basidiomycetes, the group to which the toadstools belong.

The lichens are found growing in a great variety of different places, and often they are seen in situations where neither partner could exist alone—a tribute to the effectiveness of the partnership. *Xanthoria parietina*, for instance, makes great orange splashes of colour on bare rocks near the sea. Many lichens grow on tree trunks or on the roofs of old cottages. One thing that no lichen can stand is pollution of the air with smoke from industry; you seldom find lichens growing near a large town. One exception to this is the lichen *Lecanora conizaeoides*, which appears to have a fair tolerance of pollution; in the neighbourhood of towns it is often the only lichen that will grow.

Reproduction in the lichens is complicated by the fact that they are dual organisms. Most of them have means of vegetative reproduction; portions of the lichen body containing both algal and fungal elements break away from the main plant and grow on their own. The algae, being unicellular, can reproduce by the simple process of one cell splitting into two (fission), and in this way they keep up with the growth of the lichen. The fungus also produces spores by which it can reproduce itself, but not the algal partner.

Both the fungus and the alga are capable of independent existence, at least to a certain extent. The algae that are concerned in lichen partnerships are often quite common; such genera as *Chlorella* and *Trebouxia* are extremely widespread in their own right, as well as being common constituents of lichens.

Since many of the lichen algae are common, it might be thought that a spore of a lichen fungus might germinate, producing a fungal growth that would become infected with its appropriate alga and so give rise to a new lichen growth. In fact, it is extremely doubtful whether this ever occurs. Nobody has ever succeeded in reconstituting a lichen by growing the fungus in pure culture and adding to it a culture of the alga. Because it does not happen in the laboratory, there is no justification for saying that it cannot happen in Nature; but if it does, nobody has ever seen it. Some authorities even go so far as to suggest that the power of spore formation by lichen fungi is a vestigial phenomenon, the fungus 'remembering' what it used to do before it had evolved into a lichen. This may well be so.

We do not know what each partner gains from the other in the lichen symbiosis. Since the alga is photosynthetic, it is tempting to think that sugar or some other carbohydrate makes up the alga's contribution to the partnership, but this is not certain. It might be nitrogen compounds that the fungus finds it easier to get second-hand, or it might be a vitamin, necessary for its growth, that the alga synthesizes.

Many fungi can synthesize all the substances needed for their growth from suitable organic and inorganic compounds, but others cannot. Many fungi, for instance, need to be supplied with small amounts of thiamin (vitamin B_1) in order to grow well, and several other vitamins are known to be needed by certain fungi. These vitamins are manufactured by green plants and it might be the vitamins produced by the algae, rather than the sugar, that are so important for the lichen fungi.

With this in mind, many workers have tested the vitamin requirements of fungi isolated from lichens. The results have been inconclusive. Some of the fungi needed thiamin in the medium if they were to make adequate growth, while others were independent of an added supply of vitamins.

The advantage gained by the alga out of the lichen partnership is even more uncertain. It certainly gets shelter, and can live, protected by the fungus, in places where it could not survive alone. We do not know whether this is the full story. It has been suggested that ascorbic acid produced by the fungus may be used by the alga in respiration, but so far there is no evidence of this.

A feature of all lichens is their extremely slow growth. No lichen has been reported to grow more than an inch and a half a year, and most are much slower than this. Their life span is probably very great, though we do not know precisely how long. Some are believed to live for four hundred years, and the arctic species *Rhizocarpon geographicum* has been said to reach an age of four thousand five hundred years—about the same life span as one of the giant Californian redwood trees.

One of the most interesting characteristics of many lichens is the production of lichen substances, most of which appear to be peculiar to lichens. These are organic compounds of various kinds, and in exceptional cases they may amount to as much as twenty per cent of the dry weight of the lichen, though from two to five per cent is an average figure. They are generally found outside the cells, and the reason for their production is unknown.

Most of the lichen substances are colourless, but some are brightly coloured, and in the past have been used extensively as dyes. The colours of Scottish tartans owed some of their brightness, in the past, to lichen dyes, though nowadays aniline dyes are preferred. Most of the tartan dyes were derivatives of orcinol, which gave beautiful colours when treated with ammonia in the form of stale urine. The use of lichen colours for dyeing has recently been revived on a small scale in connection with home weaving.

An offshoot of the lichen dye industry which will be familiar to most people was the preparation of litmus, used in chemical laboratories as an indicator of acidity or alkalinity. This branch of the industry flourished particularly in Holland.

Lichen substances have other uses, some of them important. The ancients used them in medicines, though with doubtful efficacy, but in modern medicine antibiotics extracted from lichens have shown promising results. In particular usnic acid, obtained from species of *Usnia*, has been used successfully in combination with streptomycin in the treatment of tuberculosis.

Another old use for lichens that has recently been revived is as an ingredient of hair lotions and toilet powders. Species of *Evernia* and *Ramulina* have frequently been so employed. They are also used for fixing the scent in perfumes and toilet soaps. The most important use for lichens at the present time is as a food for reindeer in northern regions. The reindeer moss (*Cladonia rangiferina*) (Fig. 61), and the Iceland moss (*Cetraria islandica*), grow in large quantities in Lapland and similar places. They not only supply the reindeer with excellent grazing, but enough is left over for the Eskimos to harvest as winter feed.

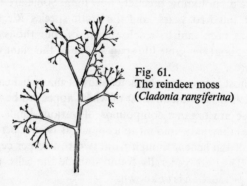

Fig. 61.
The reindeer moss
(*Cladonia rangiferina*)

Lichens have, on occasion, been eaten by man as well as animals. It is usually thought that the manna of the Bible was *Lecanora esculentia*, a lichen that grows on uplands and is frequently blown by the wind across deserts.

We have now come to the end of our study of the design of plants—or, to put it more scientifically, the adaptation of plants to their mode of life. There is no facet of their morphology, their physiology or their ecology on which the hand of natural selection has not at some time left its mark. In the long course of evolution failures have been weeded out, and the successful have been improved, so that today most plants are well adapted to the lives they lead. The process of adaptation is still going on, for the one thing that evolution cannot do is stand still.

As long as there are plants, there will be evolution, and, as the millennia go by, old designs will give way to new ones, better fitted to survive in the struggle for existence that has no end.

252

Index